SpringerBriefs in Materials

W0079195

For further volumes:
http://www.springer.com/series/10111

Md Abdul Maleque · Mohd Sapuan Salit

Materials Selection and Design

 Springer

Md Abdul Maleque
Manufacturing and Materials Engineering
International Islamic University
Kuala Lumpur
Malaysia

Mohd Sapuan Salit
Mechanical and Manufacturing Engineering
Universiti Putra Malaysia
Puchong
Malaysia

ISSN 2192-1091   ISSN 2192-1105 (electronic)
ISBN 978-981-4560-37-5   ISBN 978-981-4560-38-2 (eBook)
DOI 10.1007/978-981-4560-38-2
Springer Singapore Heidelberg New York Dordrecht London

Library of Congress Control Number: 2013954023

Printed on acid-free paper

Springer is part of Springer Science+Business Media (www.springer.com)

Preface

Writing a foreword or preface for the *Materials Selection and Design* book is indeed a pleasure. Today we are living in the world of materials and hence the applications of materials for serving mankind have increased. Of more than 1,00,000 engineering materials, we are enjoying more than 60,000 metallic materials and close to 40,000 non-metallic materials. The implication from the above is that it is highly important to focus on the selection of materials for engineering design and parts. But to really obtain very specific and most appropriate material that will lead to the prevention of unanticipated failure of the engineering structure or components, this book presents topics on the basics of materials selection and design which will give a better understanding of the selection methods and then find suitable materials for the applications. This book draws the simple and straightforward quantitative methods followed by the knowledge-based expert system approach with real and tangible case studies to show how undergraduate or postgraduate students or engineers can apply their knowledge on materials selection and design. Topics discussed in this book contain special features such as illustration, tables, and tutorial questions for easy understanding. A few published books or documents are available, hence this book will be very useful to those who use (or want to use) materials selection approach without the advantages of having had comprehensive knowledge or expertise in this materials' world.

The key feature of this book is that each chapter contains learning outcomes with step-by-step tutorials which will help the reader to learn the module/chapter quickly. A special chapter on knowledge-based expert system (KBS) has been explained for the application of KBS in materials selection. The book also elaborates numerous case studies on the materials selection of engineering components and structures, especially on how to find the most appropriate material (s) for a specific design and application.

There are six chapters in this book, with several main streams or themes such as mechanical failure of materials, design phases, materials selection processes, and knowledge-based system in materials selections. The Chap. 1 provides an overview of *Materials Selection and Design* that appears as the title of this book. Three chapters deal with the performance of materials in real applications, different stages of design, and relationship between materials' properties and design. Chapter 5 discusses elaborately the traditional or quantitative materials selection process along with several case studies. Chapter 6 is dedicated to the knowledge-based system approach in materials selection for many engineering applications. The learning outcomes and tutorial questions are included at the beginning and end of each chapter, respectively, for easy understanding and comprehensive knowledge on each particular topic of this book.

We wish to thank all the individuals who contributed directly or indirectly to the publication of this book.

Md Abdul Maleque
Mohd Sapuan Salit

Contents

Chapter 1
Overview of Materials Selection and Design

Abstract Material selection and design are integrated terms in the development of any product with a competitive cost. It is difficult to select appropriate material for any product design without having the knowledge on the importance of material selection in design. This chapter thus gives an overview on the new product development activities and some basic features of materials selection. Several new case concepts on the development of product are also introduced.

Keywords Product analysis · Activities of product development · Product life cycle · Researching the market · Recycling the materials

Learning Outcomes

After learning this chapter students should be able to do the following:

Explain the importance of materials selection and design
Identify the activities for a new product development
Basic features of materials selection

1.1 Introduction

The most important requisites for the development of a satisfactory product at a competitive cost is making sound economic choices of engineering design, materials, and manufacturing processes. It has been reported that there are more than 60,000 currently useful metallic alloys and probably close to that number of non-metallic engineering materials like plastics, ceramics and glasses, composite materials, and semiconductor (Farag 1997). This large number of materials and manufacturing processes available to the engineers, couple with complex relationships between the different selection parameters, that often make selection process is a difficult task. If the selection process is carried out wrongly, there will

M. A. Maleque and M. S. Salit, *Materials Selection and Design*,
SpringerBriefs in Materials, DOI: 10.1007/978-981-4560-38-2_1,
© The Author(s) 2013

a risk of neglecting a possible best material for certain application. This risk can be reduced by adopting a systematic selection procedure. Materials selection is one of the important activities in developing or producing a new product or improving earlier product. The product is normally expected to satisfy a certain need, to give satisfaction to the users, and to fulfill the prevailing safety and environmental laws. A product usually starts with a concept which, if feasible, will be developed into a design, and then forms a finished product. Each engineering product has its own individual characteristics and its own sequence of development events.

Selecting the optimum combination of materials and process is not a simple task nor can be performed at one certain stage in the history of a project. It should gradually evolve during the different stages of product development. In the first stages of development of a new component, one should be aware of what product to be developed or manufactured, what does it do and how does it do. Then, the important design and material requirements should be identified followed by its secondary requirement if necessary.

1.2 The Importance of Materials Selection

The materials selection plays an important role in the manufacturing process of product especially for the new product. It will not just involve with the selection of suitable materials because the design of the product should also satisfy the technical, safety and legal requirements. It is also important to manufacture it economically, to sell it at a competitive price and to dispose it satisfactorily at the end of its useful time.

Industries dealing with materials consume a considerable amount of energy in making their products. For example, 10–15 % of the total energy used in the United States is consumed by the steel, aluminium, plastics and paper producers. Most of the energy used in such industries comes from fossil fuels and therefore, contributes to carbon dioxide emission which will lead to environmental problems. Thus, in the design, materials and process selection steps, materials and/or relevant engineer should aware of the effects of the materials used and the process involved in the environment. It is not advisable and acceptable that raw materials are simply consumed in making engineering materials which are then used and discarded. In response to these demands, it is highly recommended to use materials those are recyclable and biodegradable in order to make sure that they will help to minimize the production of waste.

In addition, during the process of materials selection, the chosen materials must have suitable properties with the application of the product produced. This is to ensure that the product can operate safely and have long life cycle. For example, in the production of gear, it must use materials that show high wear resistance and have high shear strength. This is because the gears will be exposed to the friction thereby producing heat. In addition, by selecting materials that have low density, it will help to produce the product with lighter weight. This in turn will automatically help to reduce the weight of the component or part with better performance. From materials selection process, it will also help to choose the minimum cost of raw materials those have suitable properties for the product which in turn will not only reduce the actual manufacturing cost but help to produce product with lower selling price. In the main design phases, materials selection is included as part and parcel of the design activities. As a matter of fact, the main elements of design are: identification of the problem, functional requirements, system definition, concept development, materials and process selection, evaluation of the expected performance of the design, detailed design, creation of detailed drawing, preparation of bill of materials and fabrication. In a nutshell, for the development of an efficient, safe, quality and satisfactory product, material selections and design play an important role.

1.3 Relation to Design

The selection of the correct material for a design is a key step in the process because it can improve the service performance and is the crucial decision that links with a working design. Materials and manufacturing process that convert the material into a useful part underpin all of engineering design. There is a direct relationship between material selection and design especially the design configuration of the part. Therefore, it is important to deal these two (materials selection and design) together. Design engineer should identify application requirements such as mechanical, thermal, environmental, chemical etc. At the same time, it is also important to remember that this relationship is only at the primary level but is an integral part of the design process. A good example is that, a product consists of components which in turn made out of one or more parts whereby simple part consists of only one material. In that case, material selection is performed for simple parts. When several parts in a component for some reasons are to be made out of the same material, this is performed by adding this as a separate requirement followed by the analysis of the parts together with the simple part. Therefore, materials selection and design should fulfill all the requirements together and suit each other in order to have excellent and reliable product.

1.4 Product Analysis

Thousands of different products are being used every day, from telephones to bikes and drinks cans to washing machines. However, it is vital to think how they work or the way they are made. Every product is designed in a particular way and as such, product analysis enables to understand the important materials, processing, economic and aesthetic decisions which are required before any product can be manufactured. An understanding of these decisions might assist in designing and making the part completely.

The first task in product analysis is to become familiar with the product! What does it do? How does it do it? What does it look like? All these questions, and more, need to be asked before a product can be analyzed. As well as considering the obvious mechanical and other requirements, it is also important to consider the **ergonomics**, how the design has been made **user-friendly** and **marketing** issues which will provide an impact on the later design decisions.

Let's take an example of a bike:

- What is the function of a bicycle?
- How does the function depend on the type of bike (e.g., racing, or about-town, or child's bike)?
- How is it made to be easily maintained?
- What should it cost?
- What should it look like (colours etc.)?
- How has it been made comfortable to ride?
- How do the mechanical bits work and interact?

If this exercise is done for various products, certainly one can ascertain something interesting…

Systems and components

There are two main types of product—those that only have one component (e.g., a spatula) and those that have many components (e.g., a bike). Products with many components are called *systems*. In product analysis, it is preferably to start with the whole system. However, to understand various materials and processes it is required to 'pull it apart' and think about each component as well and finally, analyse the function in more detail and draft a design specification.

Design questions and specification

To build a design specification, it is required to ask the following questions:

- What are the requirements on each part (electrical, mechanical, aesthetic, ergonomic, etc.)?
- What is the function of each component, and how do they work?
- What is each part made of and why?
- How many of each part are going to be made?
- What manufacturing methods can be used to make each part and why?
- Are there alternative materials or designs in use and can propose for improvement?

These are only general questions, to act as a guide, however sometimes need to think of the appropriate questions for the products and components. For a drinks container, a design specification would look something like:

- provide a leak free environment for storing liquid
- comply with food standards and protect the liquid from health hazards

- for fizzy drinks, withstand internal pressurization and prevent escape of bubbles
- provide an aesthetically pleasing view or image of the product
- create a brand identity (if possible)
- be easy to open
- be easy to store and transport
- make it cheaper.

Once we have a specification, the next stage in the process is to understand how the materials are chosen.

Choosing the right materials

Given the specification of the requirements on each part, now can identify the material properties which will be important and an example is shown below:

Requirement	Material property
Must conduct electricity	Electrical conductivity
Must support loads without breaking	Strength
Cannot be too expensive	Cost per kg

One way of selecting the best materials would be to look up values for the important properties from the reliable sources. But this is time-consuming, and a designer may miss materials which they simply forgot to consider. A better way is to plot two material properties on a graph, so that no materials are overlooked (this kind of graph is called a materials selection chart) and has been explained in detail in Chap. 5 of this book. Once the materials have been chosen, the next step is normally to think about the processing routes.

Choosing the right process

It is all very well to choose the perfect material, but somehow we have to make something out of it as well! An important part of understanding a product is to consider how it was made—in other words what manufacturing *process is* used and why.

There are two important stages to select a suitable process:

- **Technical performance**: Is it possible to fabricate the product with the material and perform well?
- **Economics**: Is it possible to fabricate the product with lower cost?

Final remarks

Finally, it can be said that product analysis is important before any new product development and serve the followings:

- Think about the design from an ergonomic and functional viewpoint.
- Decide on the materials to fulfill the performance requirements.
- Choose a suitable process that is also economic.

Whilst this approach will often work, design is really *holistic* i.e., everything matters at once. In order to have a holistic design, the product should be *performance* driven or *cost* driven. This also makes a big difference when we choose materials. In a performance product, like a tennis racquet, cost is one of the last factors that need to be considered. In a non-performance product, like a drinks bottle, cost is of primary importance—most materials will provide sufficient performance (e.g., although polymers aren't strong, they are strong enough).

1.5 Activities of Product Development

An industrial product is normally expected to satisfy a certain need, to give satisfaction to the user and to comply with the prevailing safety and environmental issues. There are numerous approaches in product development proposed by various design experts such as Pahl and Beitz (1996) and Ashby (2005). Most of these authors emphasized on the importance of consideration design in total and design process should include customers' need, market investigation, product design specification (PDS), design concept and detail design before the fabrication can be performed. In this chapter, the approach adopted is shown in Fig. 1.1.

From Fig. 1.1, it can be seen that a product development stage basically starts from the concept then develop a complete design followed by a finished product and finally recycling of the materials. In this approach, the first two important stages in design i.e., market investigation and product PDS are not reported. Therefore, to produce a new product, there is a general pattern for the various activities that accompany the introduction of the new product.

1.6 Case Examples of Product Development Stages

In this section, three case examples of the application of product development activities for new products, i.e., piston-cylinder liner, automotive brake pad and automotive fuel tank are presented.

1.6.1 Product Development Activities for Piston-Cylinder Liner: Case Study 1

Each activity of product development for new piston-cylinder is given in detail.

Activity 1: Concept development

As a first activity, it is better to select a design concept based on the present traditional combustion engine technology. In the manufacturing of new automotive piston-cylinder liner, the first activity involves a large amount of creative work

Fig. 1.1 The activities for
the development of a new
product

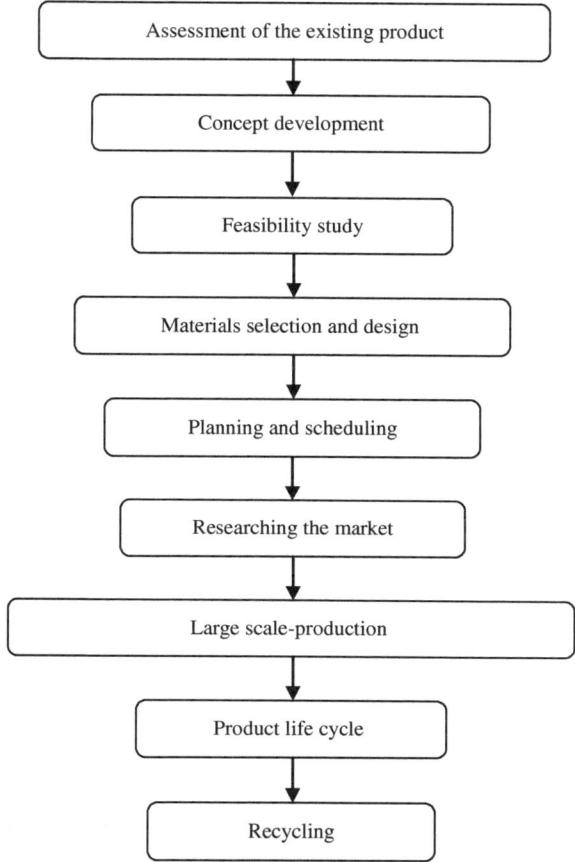

and innovation. In many cases of product development, however, an existing product is modified to suit other applications, to take the advantage of new processes and materials, to improve service performance or to comply with new legal and environment regulations. In such cases, it is an innovative part of the product development, although economic analysis would still be required.

Activity 2: Feasibility study

The second activity in the product development activity is feasibility study where sustainability issues (such as social, economical and legal) are related to the nature and functions of the new automatic piston-cylinder liner. These need to be analyzed with the emphasis on market and competition around the world. Important design features as well as main manufacturing processes and materials requirement should be broadly outlined at this stage.

Activity 3: Materials selection and design

As a result of feasibility study and comparison between the various design concepts, a final concept is selected. The design does not require a major break-
through in order to avoid the cost of new equipment and to utilize the expertise
with the current materials and processes. The activities involved estimating the
selling price of a product and selection of materials.

A workable design is developed and as a basis for more accurate estimation of
the development costs. Optimization is then performed to refine the design and to
select the optimum material and processing route. Other related departments
within the organization for examples purchasing, quality control, industrial engineering, production and marketing are consulted to determine the optimum
material procurement, manufacturing method and sale of new automotive piston-
cylinder liner.

Activity 4: Planning and scheduling

The fourth of new automotive piston-cylinder liner development is planning and
scheduling in preparation for production. Planning consists of identifying the key
activities and ordering them in sequence in which they should be performed.
Scheduling consists of putting the plan on a calendar timetable which is known as
'Gantt chart'.

Activity 5: Researching the market and pilot production

The fifth involves preparation of prototype, preliminary production and market
tests. The prototype is used to assess the performance of piston-cylinder and
feedback from customer. As a result of this development work, the piston-cylinder
liner may undergo some design or materials modifications. This also includes tool
design and tool making as well as pilot production. Even at this stage, some design
and materials modifications may have to be made in order to suit large-scale
production.

Activity 6: Large scale-production and launching

The sixth is commercial or large-scale production of the piston cylinder liner,
which is carried out concurrently with market development. The activities involve
launching covers manufacturing the product, marketing and after-sale services.
The sequence of manufacturing process is first established for part of the product
and recorded on a process sheet. The form and condition of the material as well as
tooling and production machines that will be used are also recorded on the process
sheet.

Using established standard times and labor costs for each operation, the
information in the process sheet is used to estimate the processing time and cost
for each part. The information in the process sheet is also used to estimate the
necessary stock materials, to design special tools, jigs, and fixtures to specify the
production machines and assemble lines and to plan work schedules and inventory
controls. Before starting large-scale production, a pilot batch is usually made to

test the tooling and to familiarize the production personnel with the new product and also to identify outstanding problems which could affect the efficiency of the production. The installation and maintenance instruction are required to prepare and distribute with the product. Clear installation, operation and maintenance instructions will make easier for the user to achieve the optimum performance of the product.

After sales—service for most products require regular or emergency service during their useful life. The accuracy and speed of delivering the needed service and the availability of spare parts could affect the organizational reputations and the sale volume of the product.

Activity 7: **Product life cycle**

The life cycle of the piston-cylinder liner can be prolonged by introducing mod-ification in the design to take advantage of new materials and/or new technologies. This includes an introducing a lighter piston-cylinder liner with materials that are capable of withstanding high pressure and temperature. Feedback from users to evaluate the reliability of the piston-cylinder liner and its effectiveness in per-forming the intended functions is useful in determining future modifications or developments. The availability of maintenance facilities is also important factors that influence customer satisfaction.

Activity 8: **Recycling of materials**

The last is reached when the appearance of new models or technological advances renders the piston-cylinder liner obsolescence. This causes the sale volume and production to decrease to uneconomical level, thus ending its life cycle.

1.6.2 Product Development Activities for New Automotive Brake Pad: Case Study 2

There are two brake pads placed on each brake caliper. The pad is constructed of a metal "shoe" with the lining riveted or bonded to it. The pads are mounted on the caliper; one on each side of the rotor. Traditionally, brake linings used to be made primarily of asbestos because of its heat absorbing properties and quiet operation; however, due to health risks, asbestos has been outlawed, so new materials are now being used. Brake pads wear out with use and must be replaced periodically. There are many types and qualities of pads available. The differences have to do with brake life (how long the new pads will last) and noise (how quiet they are when the brake is stepped on). Harder linings tend to last longer and stop better under heavy use but they may produce an irritating squeal when they are applied. The devel-opment of new automotive brake pad activities is explained in the following section.

Activity 1: Concept development

As a first phase of the project, it is decided to select a design concept based on a present brake pad. In this first stage, it involves a large amount of creative work and innovation. In this development of brake pad, an existing brake pad is modified to suit other applications such as, brake pad can be produced for general driving on city road or in mountainous road that requires more braking action. It is intended to take advantage of new processes and materials where the usage of free asbestos material as replacement of asbestos material and to improve its service performance or to comply with new legal and environmental regulations.

The main features of brake pad are as follows:

- It must be manufactured by asbestos free material (e.g., ceramic, Kevlar or hybrid composite)
- It must deliver maximum controlled vehicle deceleration without pitch and roll
- It must eliminate potential loss of control during hard braking and help in smooth and quiet braking performance
- It must be heat resistance
- It must have long life span
- It must have suitable friction coefficient.

Activity 2: Feasibility study

After defining the overall concept and function of the brake pad, the second activity is the feasibility study where social, economic, and legal issues related to nature and functions of the brake pad are analyzed, and questions related to the market and competition are posed. Important design features as well as the main manufacturing processes and materials requirements should be broadly outlined at this stage.

Upon this study, a market research is a necessary step in the development of the brake pad. It is essential to identify competing products and the mechanism of their operation. This is an important parameter in industries where technology is moving rapidly as well as new processes and materials could greatly affect the market ability of the product. For each competing product, it is necessary to identify:

- The technical advantages and disadvantages
- The range of applications
- Market share by volume and value.

Activity 3: Materials selection and design

As a result of the feasibility study and the comparison between the various design concepts, the best design is chosen. Figure 1.2 shows a typical brake pad design. In this process, cost as well as the quality of the brake pad produced need to be considered in order to determine the best design. In the development of the brake

Fig. 1.2 Brake pad design

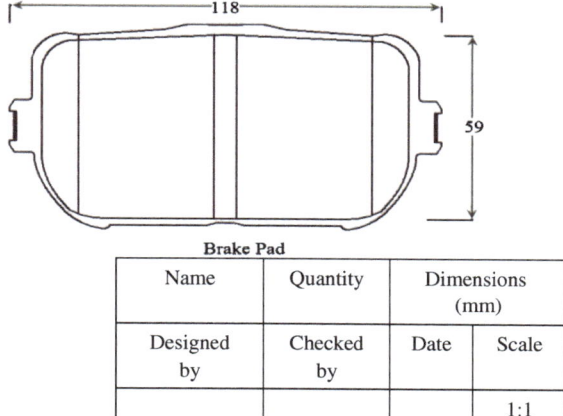

Brake Pad		
Name	Quantity	Dimensions (mm)
Designed by	Checked by	Date	Scale
			1:1

pad, recent technological advancements of processing brake pads are laser shaping and laser burnishing.Therefore, it is essential to choose the best processing method that is able to deliver the entire desired product function and at the same time focusing on cost and quality. Optimization is then performed to refine the design and to select the optimum material and processing route.

Activity 4: **Planning and scheduling**

The fourth activity of this development is planning and scheduling. This activity consists of identifying the key activities and ordering them in sequence in which they should be performed. Scheduling consists of putting the plan on a calendar time table. One example of the schedule is by using Gantt chart as shown in Fig. 1.3.

Month / Activities	1	2	3	4	5	6	7	8	9
Purchasing raw material	██	██							
Equipment installation			██						
Prototype production				██					
Modification					██	██			
Large scale production							██	██	██
Modification and development									██

Fig. 1.3 Example of Gantt chart for the development of brake pad

Table 1.1 Example of brake pad performance check list

Requirements	Prototype 1	Prototype 2	Prototype 3	Competitor's
Noise generated				
Efficiency at high temperature				
Friction coefficient				
Dust generated				

Activity 5: Researching the market

This activity involves preparation of prototype, preliminary production, and market test. The brake pad prototype is manufactured to measure the performance, braking test as well as customer reaction. Due to this process, the brake pad may undergo some design or material modification. Table 1.1 shows an example of brake pad performance check list.

Activity 6: Large scale production and launching

The sixth activity is large scale production of the brake pad which is carried out concurrently with the market development. Feedback from the users to evaluate the reliability and effectiveness in performing the desired function is useful in determining the future modification and development of this product.

Activity 7: Product life cycle

The life cycle and quality of the brake pad can be prolonged by introducing modification in the design to take advantage of new material and new technology. Figure 1.4 shows the profit and loss for the product life cycle of brake pad and it clearly illustrates that heavy costs (research and development) are involved before the launch of the product. This is a negative cash flow and great effort is usually applied to minimise this expenditure.

Activity 8: Recycling and obsolescence

The eighth and final stage is reached when the appearance of new models or technological advances render the brake pad. This causes the sales volume and production to decrease to uneconomical levels, thus ending its life cycle.

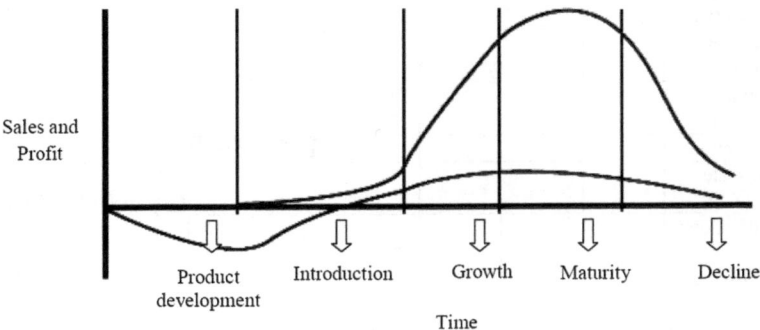

Fig. 1.4 Profit and loss for the product life cycle of automotive brake pad

1.6.3 Development of Automotive Fuel Tank: Case Study 3

A product is expected to satisfy certain needs including customer satisfaction, safety and environmental issues. The stages of product development of fuel tank are described as follows:

Activity 1: Concept development

The automotive market is global, highly competitive and rapidly changing, so the engineers must come up with the new ideas and concepts that will definitely bring the advantages. Concept development process is a blend of creative and analytical work. It is a process driven by a set of customer needs and target product specifications, which are then converted into a set of conceptual designs and potential technological solutions. These solutions represent an approximate description of form, working principles, and product features. Often, these concepts are accompanied by industrial design models and experimental prototypes that help in making final selections. In this case, the entire problems that already occurred in the previously developed fuel tank and ways to improve them need to be considered.

Activity 2: Feasibility study

A feasibility study should be performed in order to define future product development needs and how they can be met. Feasibility study of the fuel tank should cover the area of technical and engineering aspects. For example, the development targets are to produce fuel tanks that will reduce 70 % of power consumption or probably dehumidification function is added. Economic analysis, i.e., the cost of the material to produce fuel tank must be carried out. The optimum design of the fuel tank, the brackets, fittings and hoses must be studied. The pressure and temperature that the fuel tank can withstand should also be studied.

Activity 3: Materials selection and design

Selecting the optimum combination of design, materials and process is important to perform fuel tank better. The fuel tank that will be produced must satisfy all the basic safety requirements and must be aligned to the standard fuel tank. For example, it must be able to withstand cold temperature impact (especially in the Europe, the weather is very cold). The fuel tank design also should have outstanding chemical resistance.

For the material selection, the material that will be used must be easy to process; the process that will produce excellent wall thickness uniformity. Light weight fuel tank and the tank that provides gas-free expulsion of hydrazine propellant in a low gravity environment is a good design for the fuel tank. The important design requirements include:

- Weight
- Capacity
- Internal dimension
- Fluid compatibility.

Activity 4: Planning and scheduling

The purpose of planning is to identify the activities that need to be controlled. For example in the process of producing the fuel tank, the parameter that needs to be controlled includes operating pressure.

Activity 5: Prototype preparation, market testing and pilot production

Prototype must be developed and the testing of the product must be conducted. For example, complete stress and fracture mechanic analyses must be conducted to validate the tank shell in the new operating environment. Market testing is to expose the fuel tank to a small sample of the entire market to test various marketing strategies. It really helps in term of the feedbacks from the customers whether or not it satisfies their needs.

Activity 6: Large scale production, market development and sales

Large scale production cannot be done without having statistical market analysis. In order to prevent loss, the feedback from the customers is important. In this case, whether the consumers like the fuel tank produced or not and whether it fulfill the safety requirements. Large scale production will be started once the product receives positive feedback.

Activity 7: Product modification and development

After obtaining the feedback from the customers (selling it to the market or the industry) the improvement to the design will be carried out or any dissatisfaction will be analyzed. This is performed in order to fulfill the customers' needs. After fulfilling all the requirements and responding to all the feedbacks, the large scale production will be commenced.

Activity 8: Obsolescence, disposal and recycling

Finally, disposal of the rejected fuel tank and recycling of the material or making use of it in the other industries are the next activities to be performed.

1.7 Summary

For the development of an efficient, safe, quality and satisfactory product, material selections and design play an important role. There is a direct relationship between material selection and design especially the design configuration of the part. An industrial product is normally expected to satisfy a certain need, to give satisfaction to the user and to comply with the prevailing safety and environmental laws before the fabrication is performed. In order to start up a business in fabricating of new product or changing an old model it involves carrying out different activities such as concept development; feasibility study; materials selection and design; planning and scheduling; researching the market, large scale-production and launching; product life cycle; and recycling. The three case studies such as product development activities for piston-cylinder liner, development activities for new automotive brake pad and development activates of automotive fuel tank have been discussed for ease understanding and identifying the features of materials selection and design before starting the fabrication of the any new product.

1.8 Tutorial Questions

1.1 Discuss why material selection is important in design. What are the relationships between material selection and design?

1.2 A group of new Materials Engineering graduates are thinking of starting a small business in manufacturing of new educational products. What are the steps that they need to take before they can start the production of the new product?

1.3 Why does the cost of materials represent the key part of the manufacturing cost?

1.4 Explain planning and scheduling activities as a stage/involved in developing of a new product.

1.5 PHN Automotive Sdn. Bhd. in Malaysia is considering replacing cast iron by metal matrix composite in making a brake rotor for Proton car. Propose the main functional requirements and corresponding material properties that need to be considered for the above application.

1.6 What do you mean by product life cycle?

1.7 Discuss why product analysis is important in design and material selection. Explain the main aspects of product analysis.

References

Ashby MF (2005) Materials selection in mechanical design, 3rd edn. Butterworth-Heinemann, Oxford

Farag MM (1997) Materials selection for engineering design. Prentice Hall Europe, Hertfordshire

Pahl G, Beitz W (1996) In: Wallace K (ed) Engineering design: a systematic approach, 2nd edn. Springer, London

Chapter 2
Mechanical Failure of Materials

Abstract This chapter describes the major causes of mechanical failure of the engineering components or structure. Various level of materials performance is introduced. Failures due to fracture, fatigue, creep, wear and corrosion have been explained in order to understand the common mechanical failure. A case study on the failure analysis of an electrical disconnector has been presented with the recommendation to prevent the failure.

Keywords Performance level of materials · Mechanical failure · Ductile–brittle fracture · Fracture toughness · Case study

Learning Outcomes

After learning this chapter student should be able to do the following:

Suggest the factors that influence the level of performance of a material
Explain the major causes of mechanical failure
Evaluate ductile-to-brittle transition phenomenon
Justify the safe use of materials for engineering application.

2.1 Introduction

Engineering materials don't reach theoretical strength when they are tested in the laboratory. Therefore, the performance of the material in service is not same as it is expected from the material, hence, the design of a component frequently implores the engineer to minimize the possibility of failure. However, the level of performance of components in service depends on several factors such as inherent properties of materials, load or stress system, environment and maintenance. The reason for failure in engineering component can be attributed to design deficiencies, poor selection of materials, manufacturing defects, exceeding design limits and overloading, inadequate maintenance etc. Therefore, engineer should

M. A. Maleque and M. S. Salit, *Materials Selection and Design*,
SpringerBriefs in Materials, DOI: 10.1007/978-981-4560-38-2_2,
© The Author(s) 2013

Fig. 2.1 An oil tanker that fractured in a brittle manner by crack propagation around its girth (Callister 1997, 4e) (This material is reproduced with permission of John Wiley & Sons, Inc.)

anticipate and plan for possible failure prevention in advance. Figure 2.1 shows a catastrophic failure of an oil tanker that fractured in a brittle manner by crack propagation at the middle of the tanker.

2.2 Mechanical Failure

The usual causes of mechanical failure in the component or system are:

- Misuse or abuse
- Assembly errors
- Manufacturing defects
- Improper or inadequate maintenance
- Design errors or design deficiencies
- Improper material or poor selection of materials
- Improper heat treatments
- Unforeseen operating conditions
- Inadequate quality assurance
- Inadequate environmental protection/control
- Casting discontinuities.

The design of a component or structure often asks to minimize the possibility of failure. The failure of metals is a complex subject which can only be dealt with fracture or other relevant phenomenon. Therefore, it is important to understand the different types of mechanical failure i.e. fracture, fatigue, creep, corrosion, wear etc.

The general types of mechanical failure include:

- Failure by fracture due to static overload, the fracture being either brittle or ductile.
- Buckling in columns due to compressive overloading.
- Yield under static loading which then leads to misalignment or overloading on other components.
- Failure due to impact loading or thermal shock.
- Failure by fatigue fracture.
- Creep failure due to low strain rate at high temperature.
- Failure due to the combined effects of stress and corrosion.
- Failure due to excessive wear.

2.3 Failure Due to Fracture

Fracture is described in various ways depending on the behavior of material under stress upon the mechanism of fracture or even its appearance. The fracture can be classified either as ductile or brittle depending upon whether or not plastic deformation of the material before any catastrophic failure. A brief description of both types of fracture is given below.

2.3.1 Ductile Fracture

Ductile fracture is characterized by tearing of metal and significant plastic deformation. The ductile fracture may have a gray, fibrous appearance. Ductile fractures are associated with overload of the structure or large discontinuities. This type of fracture occurs due to error in design, incorrect selection of materials, improper manufacturing technique and/or handling. Figure 2.2 shows the features of ductile fracture. Ductile metals experience observable plastic deformation prior to fracture. Ductile fracture has dimpled, cup and cone fracture appearance.

Fig. 2.2 Ductile fracture in aluminum and steel after tensile testing

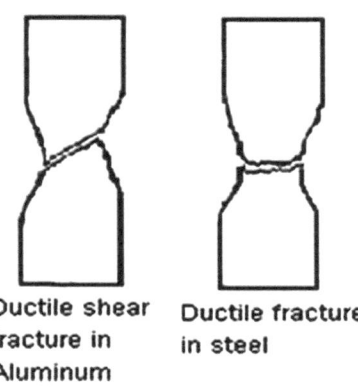

Ductile shear fracture in Aluminum   Ductile fracture in steel

The dimples can become elongated by a lateral shearing force, or if the crack is in the opening (tearing) mode. The fracture modes (dimples, cleavage, or intergranular fracture) may be seen on the fracture surface and it is possible all three modes will be present of a given fracture face.

2.3.2 Brittle Fracture

Brittle fracture is characterized by rapid crack propagation with low energy release and without significant plastic deformation. Brittle metals experience little or no plastic deformation prior to fracture. The fracture may have a bright granular appearance. The fractures are generally of the flat type and chevron patterns may be present. Materials imperfection, sharp corner or notches in the component, fatigue crack etc. Brittle fracture displays either cleavage (transgranular) or intergranular fracture. This depends upon whether the grain boundaries are stronger or weaker than the grains. This type of fracture is associated with non-metals such as glass, concrete and thermosetting plastics. In metals, brittle fracture occurs mainly when BCC and HCP crystals are present.

In polymeric material, initially the crack grows by the growth of the voids along the midpoint of the trend which then coalesce to produce a crack followed by the growth of voids ahead of the advancing crack tip. This part of the fracture surface shows as the rougher region. Prior to the material yielding and necking formation, the material is quite likely to begin to show a cloudy appearance. This is due to small voids being produced within the material. Ceramics are brittle materials, whether glassy or crystalline. Typically fractured ceramic shows around the origin of the crack a mirror-like region bordered by a misty region containing numerous micro cracks. In some cases, the mirror-like region may extend over the entire surface. The difference between ductile fracture and brittle fracture is shown in Table 2.1.

2.3.3 Ductile-to-Brittle Transition

The temperature at which the component works is one of the most important factors that influence the nature of the fracture. Sharp ductile-to-brittle transition (DBTT) is observed in BCC and HCP metallic materials as shown in Fig. 2.3.

Table 2.1 The difference between ductile fracture and brittle fracture

	Ductile fracture	Brittle fracture
Plastic deformation	Extensive	Little
Process flow	Slowly	Rapidly
Crack	Stable	Unstable
Warning signal	Imminent	No
Shape	Cup-and-cone	V or chevron
Strain energy	High	Less

Fig. 2.3 Ductile–brittle
transition temperature *curve*
(Callister 2005)

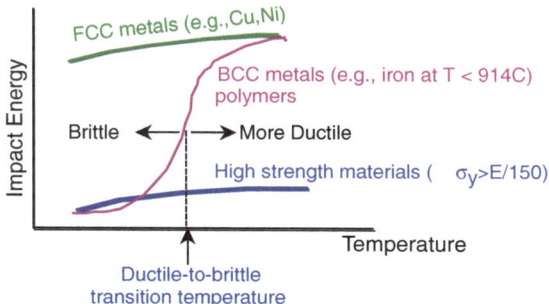

2.4 Factors Affecting the Fracture of a Material

The main factors those affect the fracture of a material are:

- Stress concentration
- Speed of loading
- Temperature
- Thermal shock.

2.4.1 Stress Concentration

In order to break a small piece of material, one way is to make a small notch in the surface of the material and then apply a force. The presence of a notch, or any sudden change in section of a piece of material, can vary significantly change the stress at which fracture occurs. The notch or sudden change in section produces what are called **stress concentrations**. They disturb the normal stress distribution and produce local co-generations of stress. The amount by which the stress is raised depends on the depth of the notch, or change in section, and the radius of the tip of the notch. The greater the depth of the notch the greater the amount by which the stress is increased. The smaller the radius of the tip of the notch the greater the amount by which the stress is increased. This increase in stress is termed the **stress concentration factor**.

A crack in a brittle material will have quite a pointed tip and hence a small radius. Such a crack thus produces a large increase in stress at its tip. One way of arresting the progress of such a crack is to drill a hole at the end of the crack to increase its radius and so reduce the stress concentration. A crack in a ductile material is less likely to lead to failure than in a brittle material because a high stress concentration at the end of a notch leads to plastic flow and so an increase in the radius of the tip of the notch. The result is then a decrease in the stress concentration.

2.4.2 Speed of Loading

Another factor which can affect the fracture of a material is the speed of loading. A sudden blow to the material may lead to fracture where the same stress applied more slowly would not. With a very high rate of application of stress there may be insufficient time for plastic deformation of a material to occur under normal conditions, a ductile material will behave in a brittle manner.

2.4.3 Temperature

The temperature of a material can affect its behavior when subject to stress. Many metals which are ductile at high temperatures are brittle at low temperatures. For example, steel may behave as a ductile material above, say, 0 °C but below that temperature it becomes brittle. The ductile–brittle transition temperature is thus of importance in determining how a material will behave in service. The transition temperature with steel is affected by the alloying elements in the steel. Manganese and nickel reduce the transition temperature. Thus for low-temperature work, a steel with these alloying elements is to be preferred. Carbon, nitrogen and phosphorus increase the transition temperature.

2.4.4 Thermal Shocks

When hot water is poured into a cold glass it causes the glass to crack which is known as thermal shock. The layer of glass in contact with the hot water tends to expand but is restrained by the colder outer layers of the glass, these layers not heating up quickly because of the poor thermal conductivity of glass. The result is the setting up of stresses which can be sufficiently high to cause failure of the brittle glass.

2.5 Griffith Crack Theory and Fracture Toughness

In 1920, Griffith advanced the theory that all materials contain small cracks but that a crack will not propagate until a particular stress is reached, the value of this stress depending on the length of the crack. Any defect (chemical, inhomogeneity, crack, dislocation, and residual stress) that exists is considered as Griffith crack, i.e. an in-homogeneity that can cause stress concentration which can be developed to failure at particular value of stress. *Fracture toughness* can be defined as being a measure of the resistance of a material to fracture, i.e. a measure of the ability of a material to resist crack propagation. *Stress intensity factor* (SIF) is another way of considering the toughness of a material in terms of intensity factor at the tip of a crack that is required for it to propagate. The parameter stress concentration factor, K_I (for mode I) is the ratio of the maximum stress in the vicinity of a notch, crack

Fig. 2.4 Fracture toughness
behavior: effect of material's
thickness

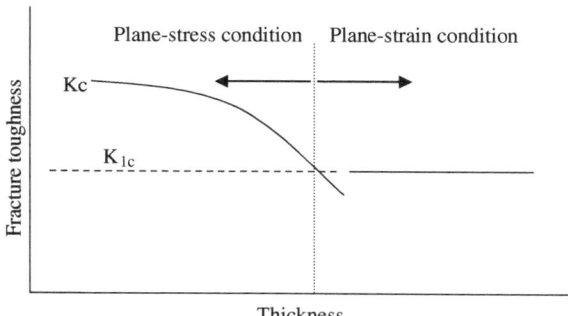

or change in section to the remotely applied stress. The stress intensity factor, K is used to determine the fracture toughness of most materials which is a measure of the concentration of stress at crack front under some consideration.

Severe fracture occurs when this SIF reaches to a critical value as denoted by K_c. The relationship between K_I and K_c is similar to the relationship between yield strength and tensile strength whereby K_c is greater than K_I. Therefore, K_c is the maximum value that can withstand by the material without any final fracture and depends on both type of materials and its thickness. The smaller the value of K_c means the less tough the material. The critical stress intensity factor K_c is a function of the material and plate thickness concerned. The thickness factor is because the form of crack propagation is influenced by the thickness of the plate. The effect of thickness on the value of the critical stress intensity factor is shown Fig. 2.4.

At large thickness, the portion of the fracture area which has sheared is very small, most of the fracture being flat and at right angles to the tensile forces. This lower limiting value of the critical stress intensity factor is called the **plane strain fracture toughness** and is denoted by K_{1c}. This factor is solely a property of the material. It is the value commonly used in design for all but the very thin sheets; it being the lowest value of the critical stress intensity factor and hence the safest value to use. The lower the value of K_{1c} means the less tough the material is assumed to be. Table 2.2 shows difference between stress intensity factor (SIF) and fracture toughness (FT).

2.5.1 Factors for Fracture Toughness

Factors those affect fracture toughness are described as follows:

2.5.1.1 Composition of the Material

Different alloy systems have different fracture toughness. Thus, for example, many aluminium alloys have lower values of plane strain toughness than steels. Within each alloy system there are, however, some alloying elements which markedly reduce toughness e.g. phosphorus and sulphur in steels.

Table 2.2 Difference between stress intensity factor and fracture toughness

Stress intensity factor	Fracture toughness
Stress intensity factor, another way of considering the toughness of a material is in terms of the intensity factor at the tip of a crack that is required for it to propagate	Fracture toughness can be defined as being a measure of the resistance of a material to fracture
Material will fail at maximum stress	As the thickness increase fracture toughness will decrease and reaches a constant value
The stress intensity factor, K is used to determine the fracture toughness of most materials which is a measure of the concentration of stress at crack front under some consideration	Fracture toughness is a measure of the ability of a material to resist crack propagation
In a flawed material, as the stress is applied the crack will propagate	Fracture toughness depends on the materials geometry and properties

2.5.1.2 Heat Treatment

Heat treatment can markedly affect the fracture toughness of a material. Thus, for example, the toughness of steel is markedly affected by changes in tempering temperature.

2.5.1.3 Service Conditions

Service conditions such as temperature, corrosive environment and fluctuating loads can all affect fracture toughness.

2.6 Failure Due to Fatigue

Metal fatigue is caused by repeated cycling of the load. It is a progressive localized damage due to fluctuating stresses and strains on the material. Metal fatigue cracks initiate and propagate in regions where the strain is most severe. Figure 2.5 shows typical S–N curve for the fatigue strength of a metal.

The process of fatigue consists of three stages:

- Initial crack formation
- Progressive crack growth across the part
- Final but sudden fracture of the remaining cross section.

2.6.1 Prevention of Fatigue Failure

The most effective method of improving fatigue performance is improvements in design. The following design guideline is effective in controlling or preventing fatigue failure:

Fig. 2.5 Schematic of S–N *curve* showing increase in fatigue life with decreasing stresses

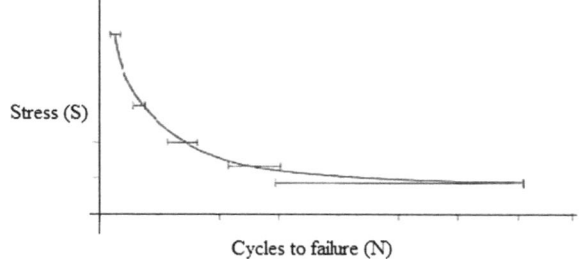

- Eliminate or reduce stress raisers by streamlining the part or component.
- Avoid sharp surface tears resulting from punching, stamping, shearing, or other processes.
- Prevent the development of surface discontinuities during processing.
- Reduce or eliminate tensile residual stresses caused by manufacturing.
- Improve the details of fabrication and fastening procedures.

2.7 Failure Due to Creep

Creep occurs under certain load at elevated temperature normally above 40 % of melting temperature of the material. Boilers, gas turbine engines, and ovens are some of the examples whereby the components experiences creep phenomenon. An understanding of high temperature materials behavior over a period of time is beneficial in evaluating failures of component due to creep. Failures involving creep are usually easy to identify due to the deformation that occurs. A typical creep rupture envelop is shown in Fig. 2.6. Failures may appear ductile or brittle manner due to creep. Cracking may be either transgranular or intergranular, if creep testing is done at a constant temperature and load, actual components may experience damage or failure at various temperatures and loading conditions.

In a creep test, a constant load is applied to a tensile specimen maintained at a constant temperature. Strain is then measured over a period of time. The slope of the curve, shown in Fig. 2.7 is the strain rate of the test during stage II or the creep rate of the material. Primary creep (known as stage I) is a period of decreasing creep rate. Primary creep is a period of primarily transient creep. During this period deformation takes place and the resistance to creep increases until stage II. Secondary creep (or stage II) is a period of approximate constant creep rate. Stage II is referred to as steady state creep. Tertiary creep (stage III) occurs when there is a reduction in cross sectional area due to necking or effective reduction in area due to internal void formation. Subsequently, increase in creep rate leading to the creep fracture or stress rupture.

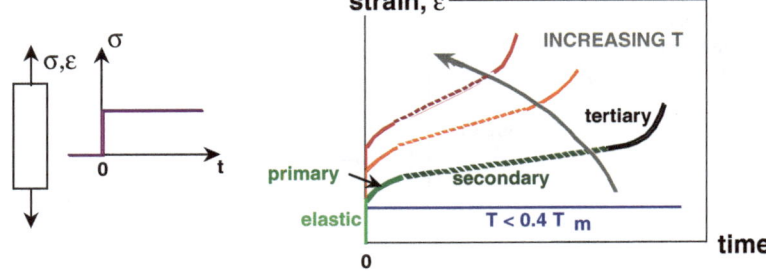

Fig. 2.6 Creep rupture envelop

Fig. 2.7 Strain rate (typical creep *curve*) of material under creep test

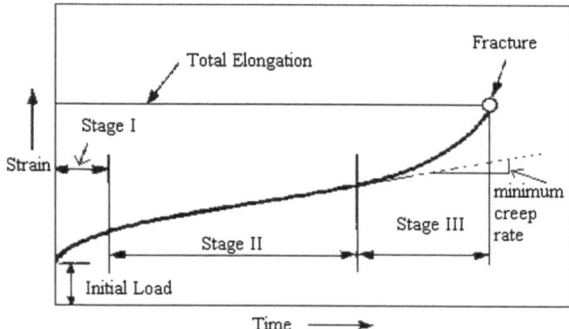

Design Problem 1

The following data apply to extruded and cold rolled nickel alloy (Nimonic 80A) at 750 °C.

Given data:			
	Young's modulus = 140 GPa		
	0.2 % proof stress = 450 MPa (minimum)		
	Elongation to fracture = 25% (short term tensile test)		
	Mean coefficient of thermal expansion (20–750 °C range) = 15.8×10^{-6}		
	The stress to cause a (plastic) creep strain in 3,000 h is		
Stress (MPa)	110	130	160
Strain (%)	0.1	0.2	0.5

Estimate the coefficient n in a power law representation between stress and strain rate. What would be the total change in length of a bar of 50 mm initial length at 20 °C, when held at a stress of 150 MPa?

Solution

Strain (%)	Stress σ (MPa)	Strain rate $\dot{\varepsilon}$, (% h^{-1})	Log($\dot{\varepsilon}$)	Log σ
0.1	110	3.33×10^{-5}	−4.48	2.04
0.2	130	6.67×10^{-5}	−4.18	2.11
0.5	160	1.67×10^{-4}	−3.78	2.20

The creep rate is related to stress by $\dot{\varepsilon} = A\sigma^n \rightarrow \log(\dot{\varepsilon}) = \log(A) + n\log(\sigma)$. The slope of the plot in Fig. 2.8 provides n = 4.3.

(a) Raising temperature to 750 °C, Thermal strain = $15.8 \times 10^{-6} \times (750 - 20) = 1.15$ %

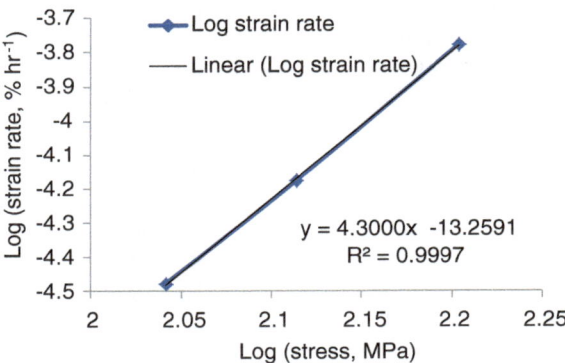

Fig. 2.8 Log strain rate versus log stress

Applying stress of 150 MPa and using $\sigma/E = \varepsilon \rightarrow$ Elastic strain = $150/140 \times 10^3 = 0.1\%$

Increase in total strain = Thermal + Elastic component of strains = $1.15 + 0.1 = 1.25\%$

For a 50 mm long bar, the extension = 0.63 mm.

(b)
$$\text{For stress} = 150 \text{ MPa} \quad Log(\sigma) = \log(150) = 2.18$$

Using graph or the linear regression $y = 4.3x - 13.2591$, $Log(\dot{\varepsilon}) = -3.9 \rightarrow \dot{\varepsilon} = 1.26 \times 10^{-4}\% \, h^{-1}$

After period of 3,000 h, material will creep and $\varepsilon = 1.26 \times 10^{-4} \times 3{,}000 = 0.38\%$

New total strain = $1.25 + 0.38 = 1.63\%$

Extension = 0.82 mm

Design Problem 2 (Creep Life estimation)

The creep rupture properties of nickel alloy (Nimonic 105) are shown in Fig. 2.9. Using Fig. 2.9, estimate the maximum operating temperature of a gas turbine blade made out of this material which is to withstand a stress of 150 MPa for a duration of 10,000 h.

What would be the new design life if the turbine engine ran 40 °C hotter?

Solution:

Larson–Miller Parameter = 27.5 (when $\sigma = 150$ MPa)

T(20 + log t)/1,000 = 27.5

T(20 + log 10,000)/1,000 = 27.5

T = 1,146 K = 873 °C

New design life if operating T goes up by 40 °C

Fig. 2.9 Stress versus
Larson–Miller parameters
graph

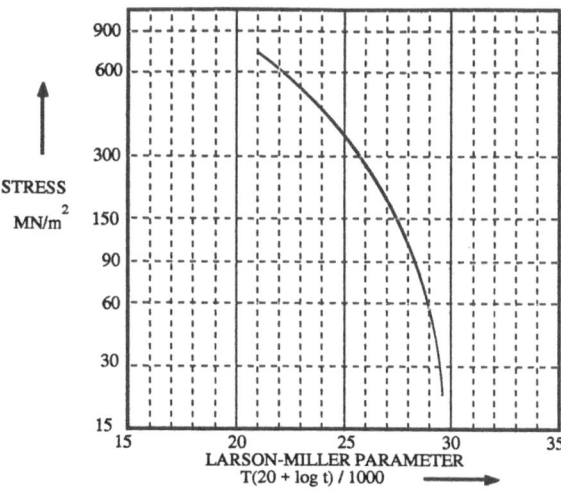

where T in °K, t is Rupture Life in hours.

T = 1146 + 40 = 1186 K
1186(20 + log t)/1000 = 27.5
20 + log t = 23.2
t = 1539 h

2.8 Failure Due to Corrosion

Corrosion of metallic materials occurs in a number of forms which differ in appearance. Failure due to corrosion is a major safety and economic concern. Several types of corrosion are encountered in metallic materials, among those: general corrosion, galvanic corrosion, crevice corrosion, pitting, intergranular, stress corrosion etc. This can be controlled using galvanic protection, corrosion inhibitors, materials selection, protective coating and observing some design rules.

   Corrosion is chemically induced damage to a material that results in deterioration of the material and its properties. This may result in failure of the component. Several factors should be considered during a failure analysis to determine the effect of corrosion in a failure. Examples are listed below:

- Type of corrosion
- Corrosion rate
- The extent of the corrosion
- Interaction between corrosion and other failure mechanisms.

As the corrosion is a normal and natural process it can seldom be totally prevented, but it can be minimized or controlled by proper selection of material, design, coatings, and occasionally by changing the environment. Various types of metallic and nonmetallic coatings are regularly used to protect metal parts from corrosion.

2.9 Failure Due to Wear

Wear may be defined as damage to a solid surface caused by the removal or displacement of material by the mechanical action of a contacting solid, liquid, or gas. It may cause significant surface damage and the damage is usually thought of as gradual deterioration. Types of wear: abrasive and erosive wear, surface fatigue, corrosive wear, fretting etc. The main feature in wear failure:

- Removal of material and reduction of dimension as a mechanical action
- Wear takes place as a result of plastic deformation and detachment of materials over a period of time.

Adhesive wear has been commonly identified by the terms galling, or seizing. Abrasive wear, or abrasion, is caused by the displacement of material from a solid surface due to hard particles or protuberances sliding along the surface. Erosion, or erosive wear, is the loss of material from a solid surface due to relative motion in contact with a lubricant that contains solid particles. More than one mechanism can be responsible for the wear observed on a particular part.

2.10 Failure Analysis of an Electric Disconnector: Case Study

2.10.1 Introduction

This section will describe a case study result on the failure analysis of an electric power station disconnector (Maleque and Masjuki 1997). At the end of this section a recommendation is made to overcome the catastrophic failure of the component. At the initial investigation it was found that the fractured disconnector for a 500 kV substation (Fig. 2.10a) was failed during installation that cause of failure is unknown. Therefore, thorough destructive examinations were performed to elucidate the causes of the failure. An inspection was conducted for the evidence of failure on site which tells nothing promising about what could have caused the current failure of the Disconnector switch. However, the following features were observed:

- Detail specification of the break disconnector
- Driving mechanism of the disconnector
- Installation procedure of the disconnector.

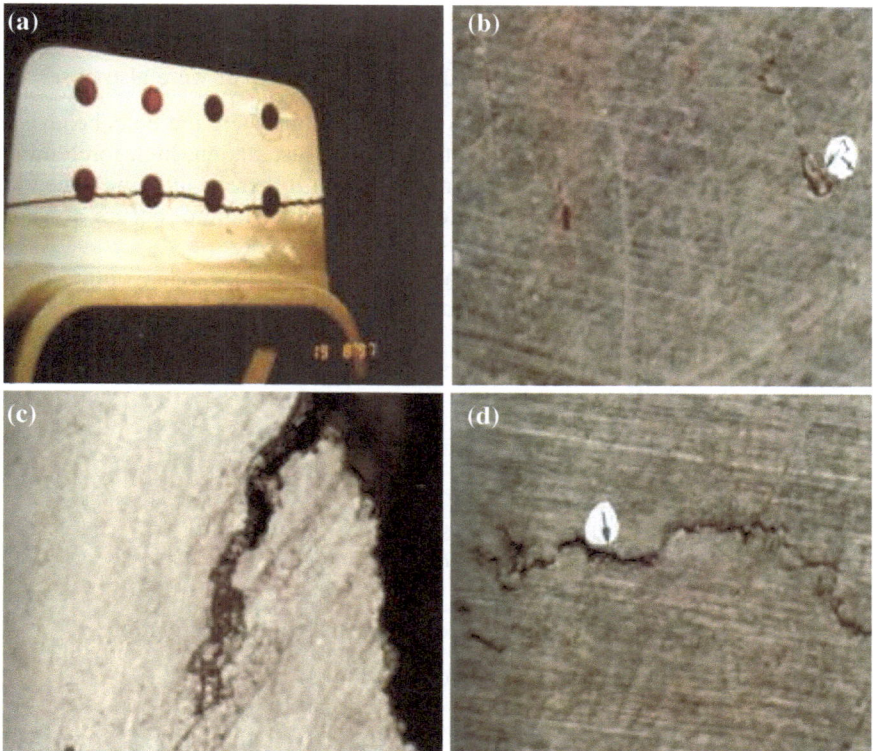

Fig. 2.10 a Fractured part of disconnector. **b** Crack pattern at the end of the fractured surface. **c** Extensive cracking at the bolt area. **d** Crack propagation at the inferior

2.10.2 Scope of Analysis

- Analysis of the failed components
- Determination of the cause and mode of failure
- Recommendation for corrective measure.

2.10.3 Visual Examination

The fractured part of the disconnector was cleaned properly and dye penetrant was applied. Cracks were found in various locations as follows:

- At the edge of the broken part as shown in Fig. 2.10a. In Fig. 2.10b, few cracks can be seen which were very close to the fractured surface, having extensive cracking.

- A wide and long crack nearby the fractured surface were found (as shown in Fig. 2.10c, d).
- Some voids or pored radiating from the surface of the part were also observed.

The nuts bolted with the hole had an interference fit where the bolts behave as integral parts. However, during installation or after installation, high force might acted on the component (such as base or blade) even if there is little movement or misalignment of the bushbar. Therefore, fracture or failure occurs towards downward direction.

From the nature and distribution of the cracks, and the appearance of the cracks surface of the disconnector, it can be suggested that:

- no plastic deformation
- rupture is downward
- surface is porous
- misalignment or mishandling of the component.

2.10.4 Metallographic Examination

The following sections were prepared for microstructural investigation:

- cross section of the blade of the disconnector
- two sections from the vicinity of the fractured part.

Before metallurgical examination, the specimens were polished and etched according to standard procedure and the microstructures were observed under optical microscope. In the photograph of the component (refer to Fig. 2.11) it can

Fig. 2.11 Optical micrograph of materials. Showing α-aluminum dendrites, acicular silicon and primary silicon plates ($\times 100$)

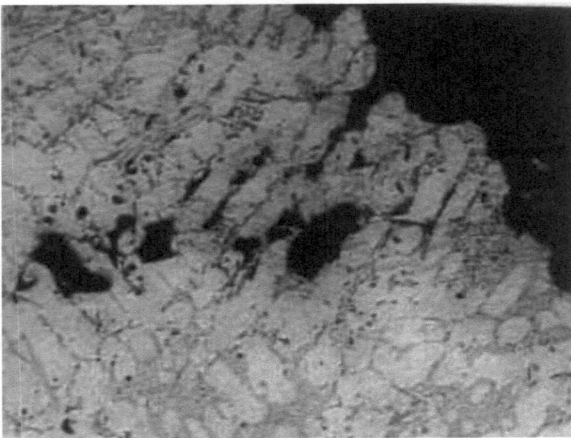

be seen that the structure consist of α-aluminum dendrites, acicular silicon and primary silicon plates. Few inclusions and voids were noticed from the microstructure of the specimen which causes inferior mechanical properties of the disconnector material.

2.10.5 Mechanical Properties

2.10.5.1 Hardness Test Result

Tests were carried out using 10 kg_{fv} and 5 kg_f load. The hardness value of the material, close to the fractured edge, was around 40 HV_5. However at most of the places, the hardness was about 72 HV_{10}.

The microhardness test was carried out on polished surface specimen. The test result is shown in Table 2.3. For the area near to the fractured layer, its hardness is lower and the hardness increased as it when goes towards bulk area, at a distance around 4 mm from the fractured edge. Almost constant microhardness value was obtained away from the edge.

2.10.5.2 Tensile Properties

The tensile properties of the Disconnector switch are given in Table 2.4. The test was done according to BS18 (1987). From the test result it can be seen that some of the parameters comply with requirements given in Aluminium Standard and Data Hand Book (1984). The percentage elongation was about 4 which is quite below the requirements. This is possibly because of the very brittle nature of the material. The breaking load was 9.65 kN.

2.10.5.3 Compressive Strength

The compressive test result is also shown in Table 2.4. It can be noticed that the material compressed by 22.68 % when the applied load was 28,000 kg_f. At the transverse direction no remarkable change occurred when the same amount load was applied.

Table 2.3 Microhardness test results of 500 kV disconnector material

Distance from edge (mm)	Load (g_f)	VHN	Average VHN
1	100	67.07, 72.1, 67.07	68.75
2	100	74.04, 73.19	73.56
3	100	75.68, 68.58	72.13
4	100	78.84, 75.68	77.26

Table 2.4 Mechanical properties of 500 kV disconnector material

Specimen	Breaking load (kN)	Proof stress (0.2 %)	Tensile strength (MPa)	Elongation (%)	Reduction of area (%)	Young's modulus (MPa)	Compressive test	Charpy impact test (J)
1	9.59	65.4	121.54	4.06	1.72	5899.14	24.18	2.95
2	10.11	62.11	127.96	4.12	1.22	7673.42	21.18	3.03
3	9.25	56.34	119.74	3.82	2.15	6716.95	22.04	3.03
Average	9.65	61.28	123.08	4.00	1.70	6763.17	22.47	3.00

Fig. 2.12 Charpy fracture surface showing the shiny, granular surface which is the characteristics of brittle fracture

2.10.5.4 Fracture Surface

The charpy V-notch Impact Energy test results are shown in Table 2.4 (at the last colum). The average value of the change in potential energy was 3.00 J. From Table 2.4, it is obvious that the Disconnector head is below capacity of the absorbed energy as far as the impact energy is concerned.

The fracture surface was very shiny, having granural appearance (refer to Fig. 2.12). However, the resulting fracture surface was relatively flat without large undulation or gross irregularities.

2.10.5.5 Chemical Composition

From chemical analysis test the material of the Disconnector seemed to be AlSiMg alloy.

The analysis result is given in Table 2.5 and was obtained from optical emission spectrometry.

Table 2.5 Microhardness test results of 500 kV disconnector material

Element	Analysis (%)
Mg	0.55
Si	9.63
Fe	0.37
Cu	0.04
Mn	0.14
Zn	0.06
Al	Remainder

From Table 2.5, it can be seen that all the chemical contents are within specification except Si. The Si percentage is quite high (9.63 %) for this type of
material and seemed to be decreased the ductility of the matrix and thus, enhanced
the brittle characteristics of the material.

2.10.6 Discussion on the Findings

2.10.6.1 Mode of Failure

- failure of the disconnector occurred due to brittle fracture
- the major cracks on the vicinity of the fracture show that they were formed
  during installation and were not new cracks.

2.10.6.2 Contributory Factor

The macro and micro examination tests showed that some inclusions and voids are
distributed throughout the matrix which seemed to be flattened. The macroscopic
examinations also shows that the color of at the fracture surface and around it is
shiny and having granular appearance. The fracture surface seems to be flat
without large undulation or gross irregularities which show that brittle fracture had
occurred. This is probably because of high percentage of silicon content in the
alloy. The high amount of Si resulted in the aluminium wrought alloy becoming
brittle.

Many cracks were found at the edge of the bolt which is an undesirable feature
because it elevated the local stress level and might initiate and propagate crack.
The presence of indentation mark closed to the fracture surface as well as edge of
the bolt resulting from misalignment seems to be one of the probable cause of
failure. Fracture occurred with no plastic deformation and proceeded along crys-
tallographic planes.

2.10.6.3 Conclusion and Recommendation

The disconnector had failed by brittle fracture due to insufficient impact energy
due to installation. The hardness of tensile properties and chemical composition
(except silicon) are within specification. However, the Si percentage is quite high
and contributes significantly to the brittle nature of the Disconnector material. The
presence of inclusions, indentation mark at the edge of the bolt as aggravate the
mechanical strength as well stress level at the point of fracture.

It is recommended that the percentage of Si of the alloy be reduced as it
promotes brittleness of the material. Adopting some metallurgical strategy is
important in order to avert or rather reduce brittleness of the material. It is also

suggested to handle the disconnector and other supporting components carefully during installation and refer to maintenance manual closely.

2.11 Summary

Engineering materials don't reach theoretical strength when they are tested in the laboratory. The usual causes of failure of engineering components can be attributed to: design deficiencies, poor selection of materials, manufacturing defects, exceeding design limits and overloading and inadequate maintenance. Flaws produce stress concentrations that cause premature failure in the component. Sharp corners in generally produce large stress concentrations leading to premature failure. Creep failure depends on both temperature and stress.

2.12 Tutorial Questions

2.1. What are the main factors that influence the level of performance of a part or component? What are the causes of failure of engineering components?

2.2. Explain the difference between stress intensity factor and fracture toughness.

2.3. Draw and explain the effect of thickness on fracture toughness behavior of materials.

2.4. Define and show both fatigue limit and fatigue strength using S–N diagrams.

2.5. Write down the common types of mechanical failures that encountered in engineering components or structures.

2.6. List down the differences between ductile and brittle fracture. Explain the ductile-to-brittle phenomenon. Support your answer with suitable diagram.

2.7. Ti–6Al–4V and aluminium 7075 alloys are widely used in making lightweight engineering structures. The fracture toughness of Ti-6Al-4 V and aluminium 7075 alloys are 55 MPa $m^{1/2}$ and 24 MPa $m^{1/2}$ respectively. The NDT equipment can only detect flaws larger than 3 mm in length. For the design of a structure that is subjected to a stress of 400 MPa,

(1) Calculate the critical crack length of both materials.
(2) Make a comment on the safe use of material for the structural applications.

2.8. AISI 4340 and Maraging 300 steels are being considered for making engineering structure. The fracture toughness of AISI 4340 and Maraging 300 steels are 50 and 90 MPa $m^{1/2}$ respectively. The NDT equipment can only detect flaws larger than 3 mm in length. For the design of a structure that is subjected to a stress of 600 MPa,

(1) Calculate the critical crack length of both materials.
(2) Make a comment on the safe use of material for designing a structural component.

Fig. 2.13 Fracture of an oil tanker

2.9. Explain what is meant by fracture toughness. Explain the terms stress intensity factor K, critical stress intensity factor K_c and plane strain fracture toughness K_{1c}.

2.10. What factors can affect the values of the plane strain fracture toughness?

2.11. Secondary creep rate, where σ is stress, Q is activation energy, R is universal gas constant, T is temperature in degrees absolute, D and n are material constants. From laboratory tests on a Nickel alloy the value of n is found to be 3. The secondary creep rate is 3×10^{-10} s^{-1} at stresses of 18 and 4 MPa at temperatures of 627 and 777 °C respectively. Determine the values of D and Q. Use the equation to find the stress which will produce the same value of at a temperature of 727 °C.

2.12. Figure 2.13 shows the fracture of an oil tanker. Explain why and how such kind of fracture phenomenon occurs?

2.13. What do you mean by fracture toughness?

2.14. What are the main features of brittle material fracture surface? Discuss the ductile-to-brittle transition temperature (DBTT) with the help of diagram. Figure below shows a severe failure of the Titanic ship. What is your recommendation to overcome such kind of failure?

2.15. Explain the fatigue limit and fatigue life for safe-life fatigue of the engineering materials. Support your answer with diagrams.

References

Anon (1984) Aluminum standard and data hand book. Aluminium Association, Washington, DC

BS 18: (1987) British Standard method for tensile testing of metals

Callister WD (1997) Materials science and engineering: an Introduction, 4th edn, Wiley, NY

Callister WD (2005) Materials science and engineering: an introduction, 6th edn. Wiley, NY

Maleque MA, Masjuki HH (1997) Failure analysis of 500 kV HAPAM DISCONNECTOR report. Technical report submitted to Transmission Technology (M) Sdn Bhd., Sept 1997

Chapter 3
Design Phases

Abstract This chapter addresses the engineering design activities followed by major phases of design. It also provides the information on the design tool and material data. The design codes, specifications and standards which are essential for design and process of any product with safety, efficiency, performance and quality are also presented.

Keywords Engineering design · Major phases of design · Design tool and material data · Design codes and specifications

Learning Outcomes

After learning this chapter students should be able to do the following:

List down different engineering design activities
Identify major phases of design
Understand about the design codes, specification and standards

3.1 Introduction

Design is the process of translating an idea of products or a market need into detailed information which a product can be made. Usually, the choice of material is dedicated by the design. It deals with the physical principles, the proper functioning and the production of mechanical system. Design work is normally carried out by the design engineer in collaboration with other departments such as customer service, marketing and sales. The product development mainly strikes the balance between what the manufacturer can produce at high quality with reasonable cost and what the user wants. Whether it is the design of a product, such as vacuum cleaner or a system such as a automotive car, creativity and innovation are the key factors in achieving a successful design. A designer is responsible for the success of a product in marketplace and it is therefore, *design activity* is given high attention in an organization. It is commonly known that although design incurs

M. A. Maleque and M. S. Salit, *Materials Selection and Design*,
SpringerBriefs in Materials, DOI: 10.1007/978-981-4560-38-2_3,
© The Author(s) 2013

only 5 % of total cost of product development, however, product design influences 70 % of the total product development cost (Parsaei and Sullivan 1993). As design cannot be performed in isolation hence this concept is advocated by concurrent engineering and research activities. In the environment of stiff competition in the marketplace, manufacturer is not only making the product that can be sold but the product must be unique so that it can remain relevant in the market and can compete in the market. Due to the competitive awareness, manufacturers are forced to develop products at lower cost, higher quality and at shorter time. Design work is normally carried out by the design engineer in collaboration with other departments such as:

• Customer service, marketing, and sales
• Legal and patents
• Quality, maintenance and finance
• Safety, codes and regulations
• R&D
• Materials, manufacturing and fabrication.

   The link between design and manufacturing process for a new product is shown in Fig. 3.1.

3.2 Factors Influence Engineering Design

The most of the engineering designs are influenced by a variety of factors which is explained below:

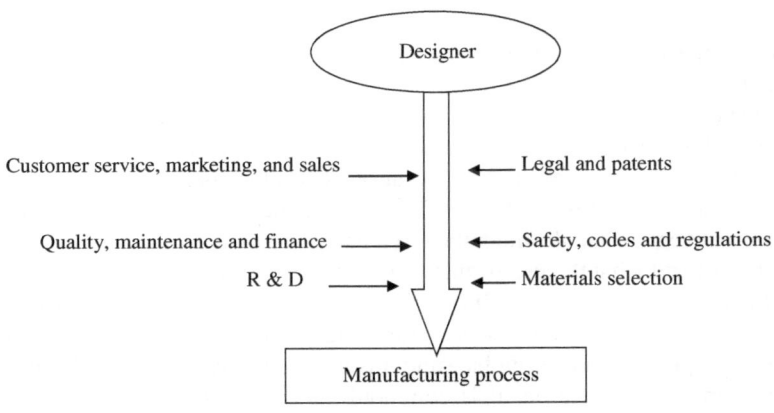

Fig. 3.1 The link between design and manufacturing process of a new product

- Product specification, such as capacity, size, weight, expected service life, safety, reliability, maintenance, human factors, ease of operation, frequency of failure, intended service environment etc.
- Design specifications, such as design codes, patents, complexity, number of parts in the system, operating loads, thermal and electrical considerations etc.
- Materials properties (or functional requirements), such as strength, ductility, toughness, stiffness, density, corrosion and wear resistance, melting point, thermal and electrical conductivity.
- Manufacturing processes, such as available manufacturing process, accuracy, surface finish, required quality etc.
- Safety and environmental requirements, such as intended service environment,
- Marketing and aesthetic parameters, such as styling and geometry.
- Economic consideration, such as raw materials cost, tool and die cost, labor and manufacturing cost, admin and other direct cost.

Based on the above factors engineering design activities can be summarized as:

- Redesign through minor modification
- Design for the new environment
- Design with no precedent.

3.3 Major Phases of Design

Design phases are usually an iterative process which involves a series of decision making steps. The main elements of the design are identification of the problem, functional requirements, system definition, concept development, materials and process selection, evaluation of the expected performance of the design, detailed design, creation of detailed drawing, preparation of bill of materials and fabrication. The design process begins with an idea and the process ends with a product that embodies the idea. Between the starting and the end point lie a set of stages such as the stages of conceptual design, preliminary design and detailed design which lead to a set of specifications which define how the product should be made. Figure 3.2 shows the block diagram of the design process where each decision establishes a framework for the next one.

A design usually passes through several steps or phases and they are:

Step 1: System requirements
Step 2: Market investigation
Step 3: Product design specification
Step 4: Conceptual design development
Step 5: Materials and process selection
Step 6: Preliminary design
Step 7: Detailed design
Step 8: Manufacturing and assembly
Step 9: Integration and testing

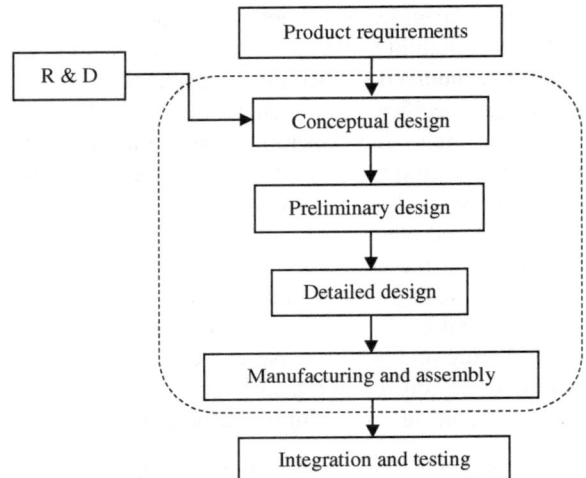

Fig. 3.2 The block diagram of major design process

A brief description of major phases of design is given below:

Step 1: Product requirements

The materials and design of a particular product generally changed over the years and will continue to do so in order to minimize the cost, process and design. However, it is necessary to find the differences between existing product and the proposed new product in terms of the above parameters. The new product should perform better than the earlier one and should have the anticipation for the last longer than ever before but with less maintenance. Therefore, the parameter that is the most important in designing a new product must be identified. The functional requirements are mainly related to the required characteristics of the product.

Step 2: Market investigation

Recognition of design need must be performed in the early stage of design. The objective of the design must be made clear to everyone involved in product development (Hyman 1998). Then, it is important to gather as many information related to design as possible. Information on the design can be found from various sources such as library sources of information (encyclopedias, handbooks, technical dictionaries, books, journals, technical reports, patents, standards, and catalogs and companies' brochures). Information from the internet can also be useful sources of information. Study of market segmentation and assessment of products of competitors among the activities of market investigation that should be performed before the design commences.

Step 3: Product design specification

In the design, all the constraints must be taken into account. Some of the design constraints such as maximum height and minimum weight can be documented in product design specification. Product design specification is also known as other terms such as product performance specifications (Nevins and Whitney 1989), or product specifications, product requirements, engineering characteristics, specifications, and technical specifications (Ulrich and Eppinger 2004).

Product design specification is a dynamic document of innovative framework (Wright 1998) that contains the objectives to be met and the constraints to be worked within. The reason why it is called a dynamic document is that the content can be changed throughout the design stage. Moreover, it is a document that can be referred to if there is disagreement between designer and customer.

Step 4: Conceptual design development

The main aim of this phase is to explore tentative general possibilities using general tools, such as calculators, formulas or backs of envelops. At the stage of conceptual design, the designer considers the options or the alternative working principles for the functions which make up the system, the ways in which sub-functions are separated or combined, and the implications of each scheme for performance or cost. Two main activities in conceptual design stage are concept generation and concept evaluation. Among the methods of concept generation include mind mapping, morphological chart, brainstorming, problem decomposition, gallery method, function analysis method, analysis of existing technical system, quality function deployment, fault tree analysis, and value analysis/value engineering (Sapuan 2010; Prasad 1996b) There is much emphasis in the recent years that during the concept development stage the product to be developed must be not only functional but also be environmentally friendly and should address the 'green' technology issue (Vallero and Brasier 2008). Concept generation is related to creativity and designer should come up with ideas using the methods mentioned earlier. Ulrich and Eppinger (2004) defined it as:

> An approximate and concise description of the technology, working principles, and form of product and how the product will satisfy the customer needs by expressing the concept as a sketch or as a rough three-dimensional model and is often accompanied by a brief textual description.

Concept evaluation is carried out to decide on the most suitable design to be further improved in the later stage of design process. Two famous methods of concept evaluation include Pugh concept selection method (Pugh 1990) and decision matrix method. Analytical hierarchy process can also be used to select the best design concept.

Pugh method initially begins with idea generation, preparation of a criteria list, picking a datum, evaluation of each alternative and improvement is made (Prasad 1996a). Decision matrix or also called weighted objective method utilizing a matrix by listing the design objectives, rank-ordering the list of objectives, assigning relative weightings to the objectives, establishing utility scores for each objective and calculating and comparing the relative utility values.

Step 5: Materials and process selection

Ashby (2005) has carried out extensive study on the role of materials in design and on materials selection. Dieter (2000) also claimed that he has put so much emphasis on the importance of material and manufacturing process in design. In the work of Pahl and Beitz (1996) materials and process selection forms is an important part of embodiment design but in this book, it is considered as a separate activity since the book itself is dealing with materials and design.

Selecting the best materials is not an easy task because various criteria that influence the selection process need to be considered. Typically, for most of the design the materials selection process begins by reviewing the materials data sheets provided by the material supplier. Without a systematic materials selection system, it is difficult to carry out a proper product design. Many computer aided systems are available materials' selection such as expert system, analytical hierarchy process and neural network. Ashby's materials data charts are very useful tools as well for this purpose (Ashby 2005) and are discussed in Chap. 5 with more details. Manufacturing process selection is another important activity in design. Dieter (2000) identified factors that determined process selection and they are cost of manufacture and life cycle cost, quantity of parts, complexity, materials, and quality of part, availability, lead time and delivery schedule.

Step 6: **Preliminary design**

The overall aim of this activity is to define and assess the main parameters of the design. It is often called embodiment design (Pahl and Beitz 1996) where it bridges the gap between the design concept and the detailed design phase (Ertas and Jones 1996). This design takes a function structure and seeks to analyze its operation at an approximate level, sizing the components, and selecting materials which will perform properly in the ranges of stress, temperature and environment and other properties that suggested by the analysis. The embodiment stage ends with a feasible layout which is passed to the detailed design stage. Embodiment means giving a concrete or discernible form to an idea and concept (Wright 1998). Pahl and Beitz (1996) defined embodiment design as:

> The part of the design process in which, starting from the principle solution or concept of a technical product, the design is developed in accordance with technical and economic criteria and in the light of further information, to the point where subsequent detail design can lead directly to production.

The above embodiment design is emphasized on the development of definitive layout of a product in this stage and proposed basic rules in embodiment design namely clarity, simplicity and safety. During this stage, many tools can be applied to consolidate the design such as computer-based techniques for synthesizing solutions and simulating the performance of various systems within the product and mathematical tools for performance analysis.

Step 7: **Detailed design**

The objective of the detailed design is to produce a documented set of performance predictions and detailed working drawings for the construction of a prototype. In detailed design stage, the specifications for each component are drawn up; critical components may be subjected to precise mechanical or thermal analysis using finite elements methods; optimization methods are applied to components and group of components to maximize performance. Pahl and Beitz (1996) considered detail design as one of the activities in embodiment design however it could be separated activity from embodiment design. Embodiment design can be replaced with detail design which could be much easier to follow and in that case detail design is one important activity total design model (Pugh 1990).

Detail design is normally related to preparation of drawings and final product specifications and the product is ready to be manufactured. Important drawings such as detail drawings, assembly drawings and installation drawings should be prepared before the product is manufactured. Specifications should include, materials requirements, test requirements, packaging requirement and maintenance provision.

Step 8: **Manufacturing and assembly**

Manufacturing is the final result of design process and manufacturing process is conducted after selection of suitable process. The manufacturing assembly process occurs before the packaging and labeling of a product. Proper design of the assembly process can simplify production and allow the manufacturer to reduce expenses and overhead. Assembly refers to putting together different parts after they are manufactured.

Step 9: **Integration and testing**

Integration and testing are the step when various parts or components of a system is integrated together and systematically tested as a whole. This is essential to meet the specified functional requirements of the final product. Therefore, the quality control or quality assurance department plays an important role here in order to ensure that the product is resilient, or flexible and capable to sustaining its expected life span as specified in the requirements without any catastrophic failure. In order to understand the above explanation clearly, a block diagram on the phases of design for designing an automotive bumper is shown in Fig. 3.3.

3.4 Design Tool and Material Data

Design tools are used to implement the set of phases those are explained earlier. Design tools are used as inputs that attached at the left of the main box of the design methodology. The tools will ease the routine aspects of each phase as the tools provide the modeling and optimization of a design. In the design tools, there are function modeling, viability studies, approximate analysis, geometric modeling, simulations methods, cost modeling, component modeling, finite element modeling, design for manufacture (DFM) and design for assembly (DFA). The function of each design tool is shown in Table 3.1.

As the design evolves, there is natural progression in the use of the tools. The progression in the use of tools such as approximate analysis and modeling at the conceptual stage, more sophisticated modeling and optimization at the embodiment stage and precise analysis at the detailed design stage.

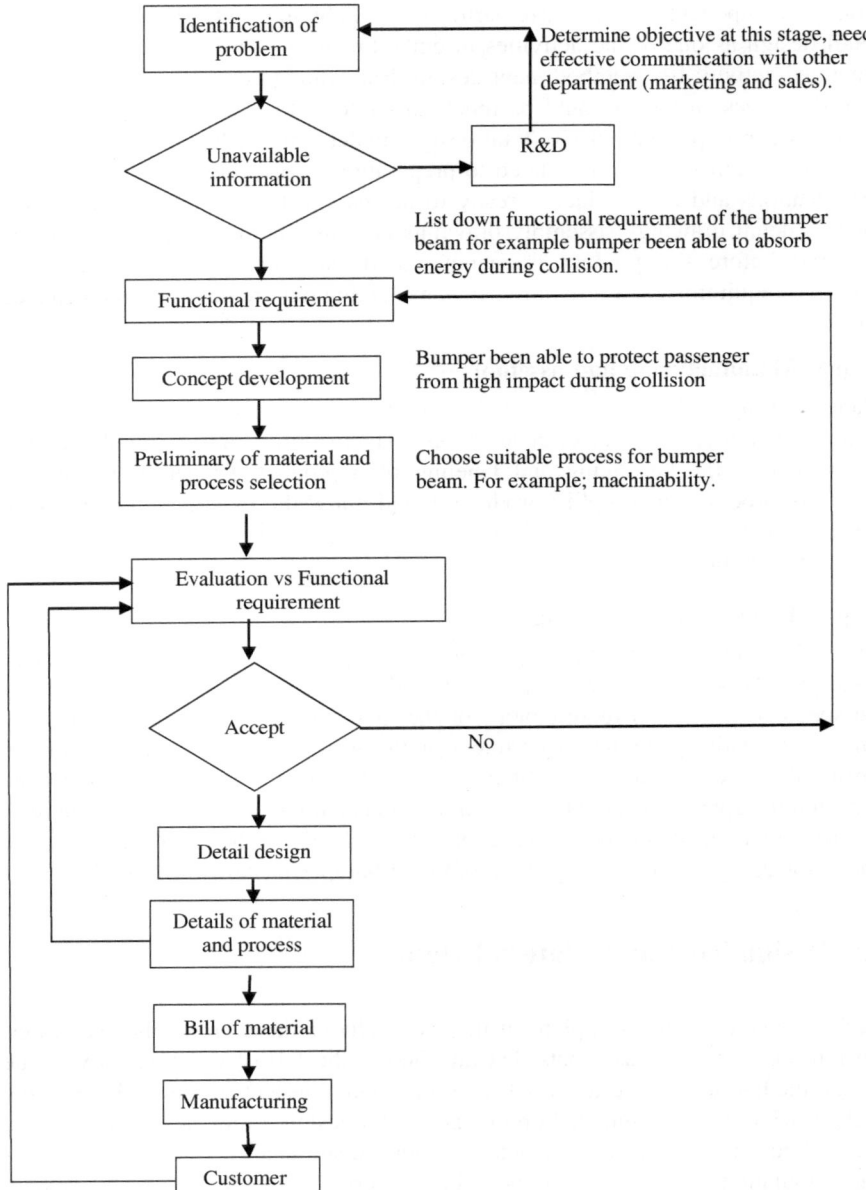

Fig. 3.3 Block diagram on the phase of design for designing and developing of an automotive bumper system

Table 3.1 The functions of design tool

Design tools	Functions
Function modelers	Suggest viable function structures
Configuration optimizers	Suggest or refine shapes
Geometric modelling	Allow visualization and create files that can be downloaded to numerically controlled prototyping and manufacturing systems
Optimization, DFM, DFA and cost estimations software	Allows manufacturing aspects to be refined
Finite element (FE) and computational fluid dynamics (CFD) packages	Allow precise mechanical and thermal analysis even when the geometry is complex and the deformations are large

At each stage of the design, materials selection is important. In the early stages, the nature of the material data needed differs greatly in the form of precision and breadth from the material data needed at the latter stages. First, at the concept stage, the widest possible data are required by the designer as to approximate property value. The widest possible data are required to make all options open or in other words, to give the freedom on the designer to choose the best material. The problem at this stage is breadth and speed of access or in other words, presenting the wide range of data to provide the designer the immense freedom in considering alternatives.

3.5 Design Reviews

Design reviews are an important aspect in the engineering design phase in order to complete the activities or steps of a design process for engineering applications. It is during the design review that all aspects of the independent and critical issues are identified and resolved for efficiency, effectiveness, and accuracy. It is crucial that all engineering applications, regardless of their size, the design is reviewed to assure that the part or product was design properly, efficient coding techniques were used, and interfaces are well understood and modified accordingly. The design review is an important process for checking the validity of design decisions and correcting errors before applications.

Multiple design reviews are recommended over the course of an application's life. There are seven basic engineering design review phases for a database application (Craig 2011):

- Conceptual design review—to validate the concept of the data and proposed application;
- Logical design review—a thorough review of all data elements, descriptions, and relationships, as well as comparison to and possibly remediation of the corporate data model;

- Physical design review—the database is reviewed in detail to ensure that all of the proper database parameter settings and other physical design choices were made, that a proper translation from logical model to physical database was made and that all denormalization decisions were formally documented;
- Organizational design review—to examine the impact of the new application/database upon the organization from a business and technological perspective;
- Application code review—a rigorous statement-by-statement review work for every statement in the application with an eye towards the accuracy and performance of the product;
- Pre-implementation design review—an overall appraisal of the system components prior to implementation;
- Post-implementation design review—formally review the application and database once it has run in production for a while to determine if the application is meeting its objectives.

3.6 Design Codes, Specification and Standards

A design code is a set of specifications for the analysis, design, manufacture and construction of a product, structure or a component. Codes of practice is set by professional groups and governments bodies in order to achieve a specified degree of safety, efficiency, performance or quality as well as a common standard of good design practice.

A standard specification is a document that describes the characteristics of a part, material, or process which is acceptable for a wide range of applications. SIRIM Berhad Malaysia is the recognised organisation that develops standards for parts, materials and processes in Malaysia. Many organizations publish codes, standards, test methods and specifications for many engineering products and design. Table 3.2 shows some of the important organizations and societies.

Name	Abbreviation
Aluminum Association, USA	AA
American Gear Manufacturing Association	AGMA
American Iron and Steel Institute	AISI
American National Standard Institute	ANSI
American Society for Metals	ASM
American Society for Testing and Materials	ASTM
International Organization for Standardization	ISO
British Standards	BS
Japan Industrial Standards	JIS
Malaysian Standard	MS

Table 3.2 Important organizations and societies

3.7 Probabilistic Approach in Design

It is very essential for the design of critical component to use a probabilistic approach in order to avoid uncertainties in the load carrying capacity (LCC) of the material which might affect by the applied load. As both are dependent on many factors it should be represented by normal distribution curves. For example, testing a tensile strength for loading 500 kN, what are the guarantees that material will sustain at that load. The measurement of sample holder, sample preparation, temperature and others are plotted whether it follow the normal statistical or not. It is good for design evaluation that avoids uncertainties might be affect by applied load.

3.8 Factor of Safety and De-Rating Factors

The factor of safety is used in design to ensure satisfactory performance. This factor is normally in the range of 1.5–5 and is used to divide into the strength of the material to obtain the allowable stress and/or the load to obtain the allowable load. The selection of the appropriate factor of safety to be used in design of components is essentially a compromise between the associated additional cost and weight and the benefit of increased safety or reliability. Generally an increased factor of safety results from a heavier component or a component made from a more exotic material and improved component design. The typical factor of safety on the basis of yield strength is shown in Table 3.3. Ductile, metallic materials tend to use the lower value while brittle materials use the higher values. The field of aerospace engineering uses generally lower design factors because the costs associated with structural weight are high for instance an aircraft with an overall

Table 3.3 Factor of safety based on the yield strength

Factor of safety	Application
1.25–1.5	Material properties known in detail. Operating conditions known in detail Loads and resultant stresses and strains known with high degree of certainty. Material test certificates, proof loading, regular inspection and maintenance. Low weight is important to design
1.5–2	Known materials with certification under reasonably constant environmental conditions, subjected to loads and stresses that can be determined using qualified design procedures. Proof tests, regular inspection and maintenance required
2–2.5	Materials obtained for reputable suppliers to relevant standards operated in normal environments and subjected to loads and stresses that can be determined using checked calculations
2.5–3	For less tried materials or for brittle materials under average conditions of environment, load and stress
3–4	For untried materials used under average conditions of environment, load and stress
3–5	Should also be used with better-known materials that are to be used in uncertain environments or subject to uncertain stresses or brittle material

Table 3.4 De-rating factors for different operating temperature

De-rating factor	
Operating temp (°F)	De-rating factor
73	1.00
80	0.88
90	0.75
100	0.62
110	0.51
120	0.40
130	0.31
140	0.22

safety factor of 5 would probably be too heavy to get off the ground. This low design factor is why aerospace parts and materials are subject to very stringent quality control and strict preventative maintenance schedules to help ensure reliability.

De-rating factors are numbers less than unity and are used to reduce materials strength values to take into account manufacturing imperfections and the expected severity of service conditions. Sometimes several de-rating factors are considered for one particular loading condition. When several de-rating factor are considered for one particular loading condition and different loading condition gives the different de-rating factor. Similarly different operating temperature will have different de-rating factors as shown in Table 3.4.

3.9 Case Study on Automotive Brake Pad

Imagine that you have been assigned to a team that will be designing and developing an automotive brake pad. As part of the project startup, your manager has asked each team member to bring a basic work plan to the next meeting. At that meeting, this work plan will be analyzed to determine the overall project time frame, cost, personnel requirements, and manufacturing process. For now, as the team member for the design team, you have been asked to bring a work plan that benefits the different phase of engineering design and includes the following information.

- Problem identification
- Functional requirements
- Detailed design
- Manufacturing process
- Challenge that you would anticipated to occur in the phase.

Please prepare the response you will bring to the meeting.

Solution

Engineering design is usually an iterative process which involves a series of decision making steps where each decision establishes the framework for the next one. In this case study, the following work plan on the phases of design is performed based on the given information.

3.9.1 Identification of the Problem

The first step is to identify the problem. The major constraints, such as cost, safety, and level of performance and the overall specifications are also defined at this stage. Of all the systems that make car, the brake system might just be the most important and is shown in Fig. 3.4. Its function determined the safety of the driver, passenger and also pedestrian.

Common problem about automotive brake pads are:

- High occurrence of brake squeals.
- High levels of brake dust generation during braking.
- Excessive rotor wear.

Ceramic-reinforced brake pads are already being used as original equipment on numerous cars. Some manufacturers claim that ceramic brake pads have reduced vibration, rotor wear and noise and dust levels compared to other brake pads. Another problem that occurs in the development of the automotive brake pads is that the brake pad is the major contributor to the pollution to the surroundings.

Fig. 3.4 A typical automotive brake system (Abdulmumin 2012)

3.9.2 Functional Requirements

If the focus is on vehicle fuel efficiency and lower emissions it means that the brakes will have to be lighter weight and not release any toxic carcinogenic substances into the atmosphere during use. This means that the choice of brake pad materials will need to be more environmentally friendly and not include toxic substances such asbestos.

Nowadays, the demands on the brake pads are such that they must:

- Maintain a sufficiently high or suitable friction coefficient with the brake disc.
- Not decompose or break down in such a way that the friction coefficient with the brake disc is compromised material is rare because this information is treated as promised, at high temperatures.
- Exhibit a stable and consistent friction coefficient with the brake disc.

And brake pads also typically comprise the following subcomponents:

- Frictional additives, which determine the frictional properties of the brake pads and comprise a mixture of abrasives and lubricants.
- Fillers, which reduce the cost and improve the manufacturability of brake pads.
- Binder, which holds components of a brake pad together.
- Reinforcing fibers, which provide mechanical strength.

Ceramics fibers can be used in place of asbestos fibres. However, further investigation should be done in order to know friction materials binders and other properties to overcome the high occurrence of brake noise, for example, the metallic pads that generates a lots of braking noise would require more fillers such as cashew and mica (noise suppressor) than barium sulphate (heat stability). Frictional additives are components added to brake friction materials in order to modify the friction coefficients as well as the wear rates. Metal sulphides appear to be better alternatives to graphite as lubricants. This is due to the current high energy braking demands in the automotive industry.

3.9.3 Detailed Design

- A good design should result in a product that performs its function efficiently and economically within the prevailing legal, social, safety, and reliability requirements.
- The main important things in order to designed the brake pad is to determine the material should be used.

Composition of the brake pad fall into four categories.

- Fibers—such as aramid, fiberglass, kevlar, stainless steel, aluminum. (can maintain the heat stability).

- Friction modifiers—graphite (adjust the friction level and fine tune the performance characteristics of the pad at specific cold and hot temperatures).
- Fillers—generally organic materials with low frictional effect.
- Resin—used to hold the material together so that it does not crumble apart.

Sealed wet friction brakes appear to be the best solution ultimately. As the brake is completely sealed from the atmosphere, there is no door of brake dust or any harmful constituents to the surroundings, thereby achieving the pinnacle of environmental friendliness. There will be no ingress of foreign particles into the brake, so braking performance inconsistencies that arise when the vehicle is stopping in rain, sand will not be an issue. The corrosion of braking components like the brake disc is also prevented.

3.9.4 Manufacturing Process

- There should be a solid plan before going to the manufacturing process. This will results to the reducing the cost and time for manufacture it. By having a solid plan also can lead us proactively choose the right component, people, tools for the jobs, to ensure a better outcome.
- In the manufacturing process, the reduction in the content of the copper is really important in order to reduce the impacts on the environment.
- After considering all the material and important properties of the materials and the types of processing can be done, then the manufacturing process will be done.
- And the inspection and test will be done on the product in order to check it's quality.

3.9.5 Challenges in the Design Phase

The main challenges that can anticipate during the design phase are:

- Safety is foremost in the brake industry. Material integrity and frictional output are the primary drivers in brake friction material or brake pad material. It is hard to maintain all the good properties in the brake pads.
- Friction material formulation is a trade-off—everyone is looking for high output, long life, no noise, no dust, no roughness, good corrosion removal, environmentally sound, very inexpensive material. This is very hard in the automotive industry. There is something we need to lose in order to get something.
- Restrictions on material usage.

3.10 Summary

Design process is a frequentative process which involves a series of decision making steps. The whole design process occurs iteratively throughout each of the design phases. There is a link between designer and the manufacturing process in order to have a good design which follows legal, social, safety and reliability requirements. The design is influenced by a variety of factors such as product specifications, design specifications, functional requirements, manufacturing process, safety and environmental requirements and cost. Design codes, specification and standard are the documents which are useful for the design in order to achieve maximum efficiency and performance of the material.

3.11 Tutorial Questions

3.1 Design is usually an iterative process which involves a series of decision making steps. In the light of this statement, explain all the phases of design those passes through an iterative process. Support your answer with suitable block diagram.

3.2 Identify the factors that affect the design of mechanical component system.

3.3 Identify different of engineering design activities that are usually performed in modifying and developing products.

3.4 Propose the main design requirements for a solid cylindrical shaft which is subjected to torsional stress.

3.5 PHN Automotive Sdn Bhd in Malaysia is thinking to replace cast iron by metal matrix composite in making a brake rotor for Proton car. Propose the main functional requirements and corresponding material properties need to be considered for the above application.

3.6 Distinguish between the factor of safety and the de-rating factor. What are the main factors that affect the value of the factor of safety?

3.7 Imagine that you have been assigned to a team that will be designing and developing an automotive piston. As part of the project startup, your manager has asked each team member to bring a basic work plan to the next meeting. At that meeting, these work plans will be analyzed to determine the overall project timeframe, costs, personnel requirements and manufacturing process.

- For now, as the team member for the design team, you have been asked to bring a work plan that identifies the different phases of design and includes the following information:

  (a) problem identification, (b) functional requirements, (c) detailed design, (d) material selection and (e) a challenge that you can anticipate would occur in the phase. Prepare the response work plan that you will bring to the meeting.

3.8 A structural member is made of steel of a tensile strength of 2,100 MPa and standard deviation on strength is 400 MPa. The static tensile stress has a mean value of 1,600 MPa and a standard deviation of 300 MPa. Find,

1. The probability of failure of the structural member.
2. Make comments on the use of the steel for structural application.

References

Abdulmumin AA (2012) Thermo-mechanical analysis of multiple-particle size reinforced SiCp aluminium matrix composite brake rotor. MEngg thesis, IIUM

Ashby MF (2005) Materials selection in mechanical design, 3rd edn. Butterworth-Heinemann Publisher, Boston

Craig SM (2011) The importance of database design reviews, Database: trends and applications (retrived from www.dbta.com)

Dieter GE (2000) Engineering design: a materials and processing approach, 3rd edn. McGraw-Hill, Boston

Ertas A, Jones JC (1996) The engineering design process, 2nd edn. Wiley, New York

Hyman B (1998) Fundamentals of engineering design, 2nd edn. Prentice Hall, Upper Saddle River

Nevins JL, Whitney DE (1989) Concurrent design of products and process. McGraw-Hill Publishing Company, New York

Pahl G, Beitz W (1996) Engineering design: a systematic approach. Springer, London

Parsaei HR, Sullivan WG (1993) Engineering: contemporary issues and modern design tools (design and manufacturing). Chapman & Hall, London

Prasad B (1996a) Concurrent engineering fundamentals, vol I. Prentice Hall PTR, Upper Saddle River

Prasad B (1996b) Concurrent engineering fundamentals, vol II. Prentice Hall PTR, Upper Saddle River

Pugh S (1990) Total design. Addison-Wesley, Reading

Sapuan SM (2010) Concurrent engineering for composites. University Putra Malaysia Press, Serdang

Ulrich KT, Eppinger SD (2004) Product design and development, 3rd edn. McGraw-Hill, Boston

Vallero D, Brasier C (2008) Sustainable design. Wiley, Hoboken

Wright I (1998) Design methods in engineering and product design. McGraw-Hill Publishing Company, Maidenhead

Chapter 4
Materials Properties and Design

Abstract The bridging between material properties and design has been presented in this chapter. The chapter begins with the interrelationship among design, materials and manufacturing processes. Design under different conditions due to complex relationship between material properties and design are explained. The importance of de-rating or modifying factors is also highlighted in this chapter.

Keywords Materials properties and design · Design under fatigue · De-rating factors · Fail-safe design · Leak-before-burst

Learning Outcomes

After learning this chapter student should be able to do the following:

Analyze the relationship between materials properties and design
Identify the de-rating factors considered for different loading conditions of the materials
Explain the modified endurance limit for the fatigue resistance of the materials

4.1 Introduction

While selecting materials for a given applications, engineers should consider a large number of material characteristics which include mechanical properties (viz. hardness, toughness yield stress etc.), physical properties (viz. density, melting point, viscosity etc.), electrical properties, manufacturing properties and find the relationship between material properties and design. Besides that user interaction aspects, such as appearance, perception and emotion are also considered for material selection (Ashby and Johnson 2002; Van Kesteran 2008). In the past, most engineers used past experiences, colour code, manufacturing code and other qualitative techniques. This selection technique is a difficult task and could take a long period of time to select a suitable candidate material. This is due to thousands

M. A. Maleque and M. S. Salit, *Materials Selection and Design*,
SpringerBriefs in Materials, DOI: 10.1007/978-981-4560-38-2_4,
© The Author(s) 2013

of materials available in the market and it should be analysed one by one. However, a successful design should take into account on the function, material properties, manufacturing process etc. The relationship between material properties and design is complex because of the behaviour of the material in the finished product can be quite different from that of the original raw material. The interrelationship among design, materials and manufacturing processing is shown in Fig. 4.1.

Material properties are linked between the basic structure and composition of the materials and service performance of the part or component. Figure 4.2 shows the journey of materials from the atomic structure up to the performance in service life. This also showed the importance of materials properties in the selection of material. One should connect the material property and manufacturing process with design as design system requires numerous material properties requirements.

4.2 Materials Properties and Design

Design of particular product or component is successful when function, materials properties and manufacturing process are taken into consideration as shown in Fig. 4.1. The relationship between material properties and design is just like the relationship between materials selection and design. Materials properties such as stiffness, elasticity, toughness, fatigue, ductility etc. are mainly determine the product quality, performance, safety and functionality which follow the design requirements. Therefore, understanding of materials properties are essential and

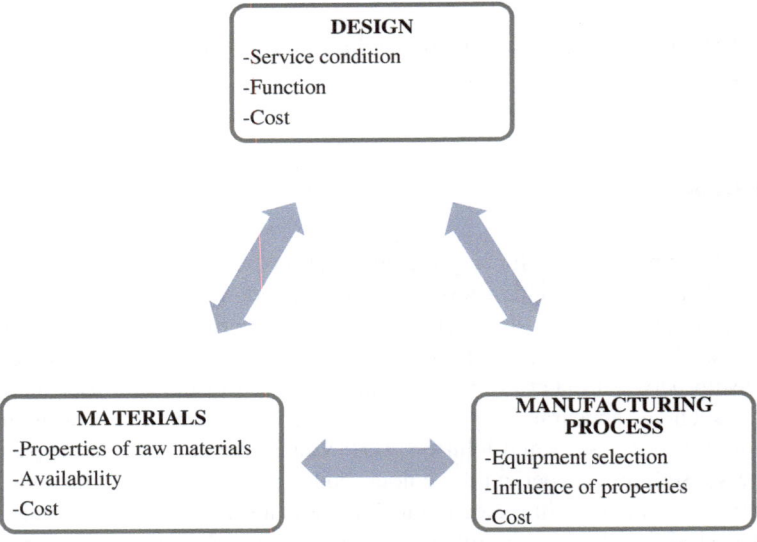

Fig. 4.1 Interrelationship among design, materials and manufacturing processing

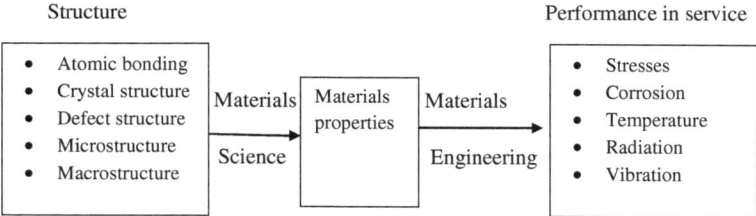

Fig. 4.2 Journey of materials from structure to performance in service

also measured them with great care in order to avoid any crack or premature failure. Another aspect is that the behavior of the material in the finished part can be quite different from that of the stock or raw material. In a nutshell, the relationship between material properties and design are not that easy as we think of Fig. 4.3 shows the direct influence of stock material properties on behavior of material in the final component as well as indirect relation to the component geometry and manufacturing process.

Several examples in relation to the above are given below:

- **Fail-safe design** requires a structure to be sufficiently damage tolerant to allow defects to be detected before developing any over (dangerous) size
- **Leak-before-burst** is a useful design criterion in case of pressure vessels and similar products. The toughness of the material used for such applications should be sufficiently high to tolerate a defect size which allows the contents to leak out before it grows catastrophically
- **Safe-life or finite-life** design assumes that the component is free from flaws, however, the stress level in certain areas is higher than the endurance limit of the material. In such case, fatigue-crack initiation is inevitable and the life of the component is estimated on the basis of the number of cycles to initiate such a crack.

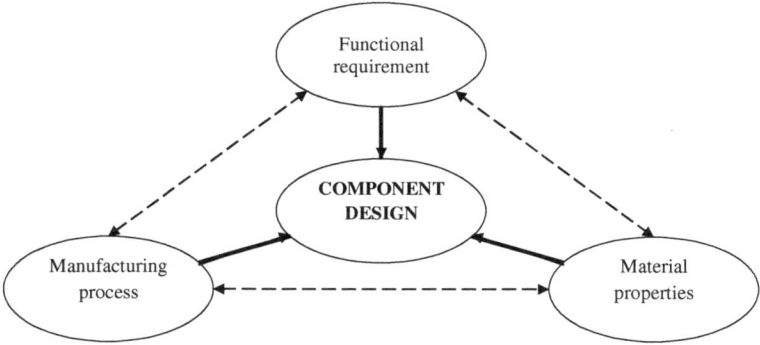

Fig. 4.3 Basic factors that affect the successful design of a component

For a successful mechanical design, a designer should take into account on the following factors:

- Function and consumer requirements
- Materials properties (both raw and finished form)
- Manufacturing process
- R&D activity
- Geometry of the design component.

There is a central problem of material selection in mechanical design as there are close interaction among functional requirement, material properties, geometry and manufacturing process which finally affect the behavior of material. The basic and anticipating factors that affect the behaviour of materials in successful component design of a component are shown in Fig. 4.3. It is clearly seen from Fig. 4.3 that the component design lies at the centre of three interacted or interrelated factors. However, the interaction among function, material properties and manufacturing process are also co-related and lie down at the heart of the material selection process. This is mainly due to the fact that materials' performance is always depending on the manufacturing process. On the other hand, function states the choice of material and shape. Process is influenced by the properties of the material such as formability, machinability, weldability and so on as it interacts with the shape of the final component. The process also influences the shape, size, precision, and cost. The interactions are two ways as specifications of shape restricts the choice of material for a particular process but equally the specification of process limits the material that can be used and the shape that can be taken. The more sophisticated the design, the tighter the specifications and the greater the interactions.

For any design of an engineering product, the functional requirements must be considered in advanced. Example of functional requirements includes strength, elastic deformation, lighter weight, component geometry etc. Materials properties are also considered properly or carefully on the basis of the essential materials requirements of the component. Example of materials properties include higher strength, lower density, better corrosion and wear resistance etc.

4.3 Design Under Different Conditions

Because of the complex relationship between design and behavior of actual materials in the component, design should be performed under different conditions such as:

- Design under static strength
- Design for axial load
- Design for torsional load
- Design for bending

- Design against fatigue
- Design under high-temperature conditions
- Design for hostile environment.

The laboratory test results (under lab condition) cannot be directly used for design purposes due to the behavior of component not only depend on loading or stress but also some other variables resulting from manufacturing imperfection, design and improper material selection. Therefore, modifying factors or de-rating factors can be employed to account for the main parameters those might affect the materials performance in service.

The de-rating factors are numbers less then unity and used to reduce materials strength value to take into account manufacturing imperfection, incorrect design and improper materials selection. Commonly used de-rating factors (parameters) in design are discussed below.

4.3.1 Surface Finish Factor

Surface finish factor is important because most machine elements are not manufactured with the same high quality finish. The values for surface finish factor are between 1 and 0.2 depending on surface finish and strength of the materials. Table 4.1 shows effect of surface finish on surface finish factor (k_a) for reinforced plastics.

4.3.2 Size Factor

Larger size component will have always lower strength compared to smaller size component due to more flaws or defects in the larger component. The general guideline for assuming size factor is (Farag 2007):

- 1.0 for component dia less than 10 mm
- 0.9 for component dia in the range of 10–50 mm
- $[(D - 0.3)/15]$, where D is assumed between 50 and 225 mm.

Table 4.1 Effect of surface finish on surface finish factor (k_a) for reinforced plastics (Farag 2007)

UTS (MPa)	Forged $R_a = 500 - 125$	Hot rolled $R_a = 250 - 65$	Machined or cold drawn $R_a = 125 - 32$	Ground $R_a = 63 - 4$	Polished $R_a < 16$
420	0.54	0.70	0.84	0.90	1.00
700	0.40	0.55	0.74	0.90	1.00
1,000	0.32	0.45	0.68	0.90	1.00
1,400	0.25	0.36	0.64	0.90	1.00

4.3.3 Reliability Factor

It is considered for the random variations of the strength of the materials. The literature value and experimental value may not same hence, it is advised to adopt reliability factor for safe design as per following guideline:

- 0.900 for 90 % reliability
- 0.81 for 99 % reliability
- 0.75 for 99.9 % reliability.

4.3.4 Operating Temperature Factor

It is essential to consider the test temperature and actual operating temperature of the component.

4.3.5 Loading Factor

It is important due to differences in loading between lab test and service. Some other loads may encounter during service of the component such as vibration, shocks, overload, transient or type of loading etc. Hence a guideline for this factor is:

- 1.0 for reverse bending test
- 0.9 axial loading condition
- 0.58 for torsional loading condition.

4.3.6 Stress Concentration Factor

This affects the strength or properties of the materials as high-strength steels are more sensitive to notches than low-strength, ductile steels.

4.3.7 Service Environment Factor

As the strength of the material changes with the changes of environment especially hostile environment, it is advisable to have the service environment factor.

4.3.8 Manufacturing Process Factor

This factor also influences the final properties of the component and is associated with heat treatment, cold working, residual stresses, protective coating etc.

4.4 Designs Against Fatigue Load

The fatigue behavior of material is usually described by means of S–N diagram which gives the number of cycles to failure (N) as a function of maximum applied alternating stress (S). However, for design purposes, laboratory fatigue test results are not applicable as the behavior of component subjected to fatigue loading does not depend only on fatigue or endurance limit but also on several other factors, such as:

• Size and shape of the component
• Type of loading
• Stress concentration
• Surface finish
• Operating temperature
• Service environment
• Method of fabrication
• Mean stress levels
• Metallurgical variables.

All the above factors should be taken into consideration as these are happened mainly due to manufacturing imperfection, design and improper material selection. In designing fatigue resistance, the following de-rating or modifying factors are considered:

$$S_e = k_a k_b k_c k_d k_e k_f k_g k_h S'_e. \qquad (4.1)$$

where,

S_e endurance limit of the material in the component
k_a surface finish factor
k_b size factor
k_c reliability factor
k_d operating temperature factor
k_e loading factor
k_f service concentration factor
k_g service environment factor
k_h manufacturing process factor
S'_e endurance limit as determined by laboratory fatigue test.

The Eq. 4.1 can be used to predict the behaviour of component or structure under fatigue condition provided that the values of different modifying factors are known.

4.5 Design Problem

A cylindrical shaft whose two-third of the mass is supported by an axle of a structural component (5,000 kg) and act as a cantilever beam. The length of the shaft is 1 m and material is heat-treated AISI 4340 steel and tensile strength is 950 MPa. The dimension of the shaft requires a gradual change in diameter to allow the fitting of bearing with it. The endurance limit and endurance ratio (Tensile strength/endurance limit) of the material are 530 MPa and 0.55 respectively. Using above information determine the diameter of the shaft which can be used in structural member.

Solution:

Assumptions

- Shaft will be finished by machining
- Expected shaft diameter is within the range of 50–225 mm
- Reliability of shaft should be high
- Shaft should be loaded in bending
- Rounded fillet should be incorporated at the changing of the diameter in order to reduce the stress concentration.

Endurance ratio of steel (ER) = 0.55
First find all possible de-rating factors:

1. K_a = surface finish factor
   Tensile strength = 950 $\approx$ 1,000 MPa, thus K_a = 0.68
2. K_b = size factor
   Assume the range of diameter (D) is between 50 and 225, so, D = 75 mm corresponds to 2.95 inch

$$K_b = 1-(D - 0.03/15)$$
$$= 1-(2.95- 0.03/15)$$
$$= 0.8$$

3. K_c = reliability factor
   As the reliability of the shaft is high, so consider 99.9 % and hence, K_c = 0.752
4. K_d = operating temperature
   K_d = 1
5. K_e = loading factor, since shaft is subjected to bending, hence, K_e = 1

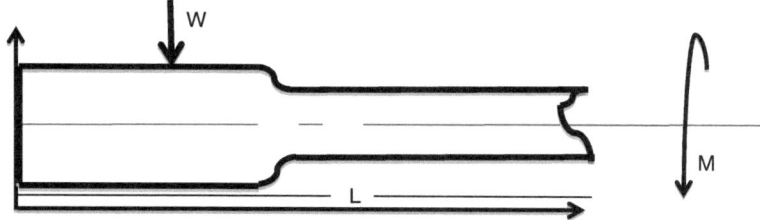

Fig. 4.4 Free body diagram of the shaft

6. K_f = stress concentration factor, as large fillet radius is required, hence,
   $K_f = 0.7$
7. K_g = service environment factor
   $K_g = 1$ (as no acid, no corrosion environment)
8. K_h = manufacturing process
   $K_h = 1$

Modified endurance limit of the shaft is (Fig. 4.4):

$$S_e = k_a k_b k_c k_d k_e k_f k_g k_h S_e'$$
$$= (0.68)\,(0.8)(0.752)(1)(1)(0.7)(1)(1)(530) = 152\,\text{MPa}.$$

From the free-body diagram of the shaft, find the load acting on the shaft:

$$R_1 = (5000 \times 9.81) \times 2/3(L/2)$$
$$= 152\,\text{MPa}$$

Apply bending moment:

$$M_t - (R_1)(1\,\text{m}) = 0$$
$$M_t = 16,352\,\text{Nm}$$

Bending stress equation for the shaft:

$$Se = 32M_t/\pi D^3$$
$$D^3 = 32\,Mt/\pi(S_e)$$
$$D = 103\,\text{mm}$$

4.5.1 Design for Automotive Intake Manifold: A Case Study

Background:

Design a car component next to the engine, such as intake manifolds made from glass reinforced plastics (Nylon 66) composite. Primary function of the intake manifold is to evenly distribute the combustion mixture or just air in a direct injection engine to each intake port in the cylinder heads. The intake manifolds made from Nylon 66 which contains 40 % glass having 200 MPa tensile strength. Using Tables 4.2 and 4.3 find the modified endurance limit of the intake manifold material which can be used safely without any catastrophic failure.

Solution:

From Table 4.2, the endurance ratio of 40 % glass fibre reinforced Nylon 66 is 0.31 and the endurance limit is 62.7 MPa. Assuming that the intake manifolds will be finished by machining, the surface finish factor k_a can be taken as 0.84, referring to Table 4.3. As the intake manifold diameter is expected to be in the range of 50–225 mm, the size factor k_b can be taken as 0.8, assuming a 3 in (75 mm) diameter. The reliability of the intake manifolds should be high and the reliability factor is taken as $k_c = 0.752$. The loading factor can be taken as $k_e = 0.58$ as the intake manifolds are loaded in torsional. By using Eq. (4.1), the endurance limit of the intake manifolds, S_e is:

$$Se = k_a k_b k_c k_d k_e k_f k_g k_h S'_e, \qquad (4.2)$$

where,

Se   endurance limit of the material in the component
k_a surface finich factor
k_b size factor
k_c reliability factor

Table 4.2 Comparison of static and fatigue strength of some engineering material

Material (Polymer matrix composite)	Tensile strength (MPa)	Endurance limit (MPa)	Endurances ratio
Reinforced Polyester with 30 % glass fibre (30 % GFRPE)	123	84	0.68
Reinforced Nylon 66 with 40 % glass fibre (40 % GFRN66)	200	62.7	0.31
Reinforced Polycarbonate with 20 % glass fibre (20 % GFRRPC)	107	34.5	0.32
Reinforced Polycarbonate with 40 % glass fibre (40 % GFRPC)	131	41.4	0.32

Table 4.3 Effect of surface finish on surface finish factor (k_a) for reinforced plastics

Tensile strength (MPa)	Forged $R_a = 500 - 125$	Hot rolled $R_a = 250 - 65$	Machined or cold drawn $R_a = 125 - 32$	Ground $R_a = 63 - 4$	Polished $R_a < 16$
420	0.54	0.70	0.84	0.90	1.00
700	0.40	0.55	0.74	0.90	1.00
1,000	0.32	0.45	0.68	0.90	1.00
1,400	0.25	0.36	0.64	0.90	1.00

k_d operating temperature factor
k_e loading factor
k_f service concentration factor
k_g service environment factor
k_h manufacturing factor
S'_e endurance limit of the material as determined by laboratory fatigue test.

Therefore,

$$S_e = 62.7 \times 0.84 \times 0.8 \times 0.752 \times 0.58 = 22.97 \text{ MPa}.$$

The value of endurance limit S_e is lower than the value of endurance limit from the Table 4.3 thus the design assumptions for the material properties and aspect design are correct and can be considered.

4.6 Summary

The relationship between material properties and design is complex because the behaviour of the material in the finished product can be quite different from that of the stock or original raw material. Because of the complex relationship between design and behavior of actual materials in the component, design should be performed under different conditions such as: design under static strength, design for axial load, torsional load, bending and fatigue separately. In designing fatigue resistance, a list of de-rating or modifying factors is needed to be taken into consideration.

4.7 Tutorial Questions

4.1 Identify and discuss the factors that affect the behavior of materials in component design? Support your answer with block diagram.
4.2 What do you mean by de-rating factors? In designing fatigue resistance, the designer should consider the modified or de-rated endurance limit. Identify the de-rating factors those are considered for the fatigue resistance of the materials. Explain why de-rating factors are important in successful design?
4.3 The two rear axle shafts for a structural application is fully loaded with a gross mass of 4,000 kg. Assume that two-thirds of the mass of truck is supported by the axles, and the construction is such that the axles can be treated as cantilever beams each of 1 m length with the load acting on its end. The shaft material is AISI 4340 steel of tensile strength 600 MPa. The shaft construction requires a change in diameter to allow the fitting of bearings. Calculate the diameter of the shaft.
Make the following assumptions:

1. The endurance ratio and endurance limit of AISI 4340 steel is 0.4 and 150 MPa respectively.
2. The expected shaft dia is within the range of 50–225 mm.

4.4 Identify the relationship between materials properties and design.

4.5 Define the factors that affect the behavior of materials in component design? Support your answer with block diagram.

4.6 The main material for the dashboard of a car is 40 % glass fibre reinforced Nylon 66 polymer composite. Tables 4.2 and 4.3 show the materials properties and surface finish factors for polymer matrix composites respectively. The dashboard diameter is expected to be in the range of 75–125 mm and is subjected to torsional loading. Other de-rating factors can be assumed accordingly based on the functional requirements of the intake manifold.
Using Tables 4.2 and 4.3, find the modified endurance limit of the 40 % glass fibre reinforced Nylon 66 polymer composite dashboard which can be used safely without any catastrophic failure.

4.7 You are required to design a component which is subjected to fatigue loading. According to design guideline, the modified endurance limit has to be considered for safe design. Identify and explain all the possible de-rating factors for the modified endurance limit in fatigue design.

References

Ashby MF, Johnson K (2002) Materials and design: the art and science of materials selection in product design. Butterworth-Hienemann, Amsterdam

Farag MM (2007) Selection of materials and manufacturing processes for engineering design. Printicee Hall, London

Van Kesteran I (2008) Product designer's information needs in materials selection. Mater Des 29:133–145

Chapter 5
Materials Selection Process

Abstract The fundamental criteria and concept of material selection are addressed in this chapter followed by the quantitative methods of material selection for many engineering design and products. A basic concept on the events of materials selection process is also presented. The quantitative methods provide the guidelines for the students and engineers to select the material easily for ant design. Several case studies on the selection of materials for automotive and structural applications are presented. The merits and demerits of different quantitative methods are highlighted.

Keywords Materials selection criteria and concept · Materials performance requirements · Ashby's material selection chart · Cost per unit property method · Digital logic method · Performance index

Learning Outcomes

After learning this chapter students should be able to do the following:

> Identify the events in materials selection
> Analyze materials performance requirements
> Develop alternative solutions for screening the materials
> Evaluate different solutions using quantitative approach
> Make decision on the optimum selection.

5.1 Introduction

Materials technology has a profound impact on evolution of human civilization. Historians have characterized evolution period as the Stone Age, Bronze Age, the Iron Age and Silicon or Synthetic material Age. Each new era was brought about by the continuing quest for even better products, a quest that is very much in evidence today. The current 'synthetic materials age' has been precipitated by people's demand for materials with superior performance characteristics, inspired

M. A. Maleque and M. S. Salit, *Materials Selection and Design*,
SpringerBriefs in Materials, DOI: 10.1007/978-981-4560-38-2_5,
© The Author(s) 2013

primarily by the quest to conquer the last frontier of space. The new class of engineered synthetic materials comprising ceramics, plastics, and composites has had a major impact upon human lifestyles in almost all areas (Ghandhi and Thompson 1992). At the beginning of the Palaeolithic period, natural objects were used which includes fibres, skin, bones and stone. Materials were selected based on the structure or structural geometry. They focused on mechanical properties such as strengths. During Stone Age, flint was used due to the structure which was better compared to bone and wood. Polycarbonate, steel, wood, iron, concrete are some examples of structural materials. As the time passed the revolution occurred, quest for better living and hence moved to functional properties. From there materials were selected based on their function or functional properties not based on structural property only. Examples of functional materials are nickel titanium, cadmium sulphide and others. Nowadays there are thousands of materials available in hands, therefore, it's more complicated process and task to select a material for certain application. Many parameters should be taken into account such as design, manufacturing process, cost etc. Physical and meta-physical aspects also should be taken into consideration (Lennart and Edwards 2003).

The main three domains are design, materials selection and manufacturing process play very important role in the development of satisfactory products. As more than 60,000 metallic materials and close to 40,000 of non-metallic materials (viz., plastics, ceramics, glasses, composites and semiconductor materials) are available to the engineer, therefore, it is a difficult task to make a selection of appropriate materials for a particular product. Risk might occur due to overlooking of a possible attractive alternative solution. Therefore, it is possible to reduce this risk by implementing a systematic selection procedure. Advantages of material selection:

- Utilization of new material or manufacturing process
- Improvement in the service performance
- Meet or compliance with new legal requirements
- Reducing cost and making the product more competitive.

Material selection can in principle be made during any stage of the life cycle of a product. The first and most important step in selecting a material from the broad range of available materials is to carefully define the requirements of the application, the physical properties required and the environment in which the material will need to perform. In the first stages of development, one should find out the fundamental criterion of material selection by answering the following questions:

- **WHY**
- **WHEN**
- **WHO** and
- **HOW**

Table 5.1 Fundamental criteria of materials selection

Fundamental criteria of materials selection

Question	Answer
Why	For maximum utilization of engineering materials
	To reduce the cost of expensive structure
	To overcome the failure
When	To develop an existing product and designs
	To adapt an existing product and design
	To create a totally new design
Who	Design engineer in collaboration with materials engineers
How	Identify the functional requirements
	Find suitable materials properties with specifications
	Find a group of materials
	Do the ranking of the materials with performance index
	Identify most appropriate materials for the application

Answering these questions will help in specifying the function of the components or the functional requirements of the component or part design. A typical example of fundamental criteria for materials selection process is shown in Table 5.1.

5.2 Events in Materials Selection Process

For materials selection, there is a small number of methods that have evolved to a position of prominence. Materials selection process is an open-ended and normally leads to several possible solutions to the same problem. This can be illustrated by the fact that similar component performing similar function, but fabricated by different manufacturers, are often made from different materials and even by different manufacturing processes. However, selecting the optimum combination of material and process is not a simple task rather gradually evolving processes during the different stages of materials selection. A typical concept (flow chart) of materials selection process is shown in Fig. 5.1.

Each materials selection decision has its own individual characteristics and sequence of events. The sequence of events in materials selection process is summarized as follows:

- Materials performance requirement
- Development of different solutions
- Application of quantitative methods for the evaluation of different solutions; and
- Decision on the optimum solution.

Fig. 5.1 A typical concept (*flow chart*) of material selection process

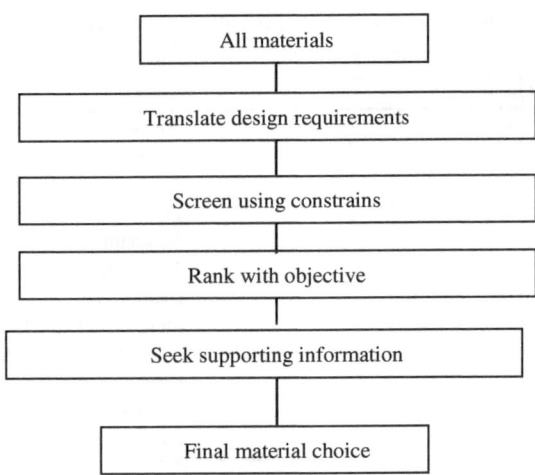

5.3 Materials Performance Requirement

Performance requirement of material is an umbrella term incorporating performance related pre-requisite and so called warranties. In the broader term, performance requirement is defined as the performance characteristics of the final product and link them to fabrication of the material. For the analysis of material performance requirements the term 'performance index' can be used. Generally, the material performance requirements can be divided into four broad categories which are functional requirements, manufacturing requirements, reliability and sustainable requirements. Cost of materials is also another vital controlling factor in evaluating materials to meet the application requirements in practical conditions maintaining the required cost limit.

5.3.1 Functional Requirements

This requirement is related to the required characteristics of a product or component. For example, if the component carries uniaxial tensile load, yield strength can be related to the load carrying capacity (LCC) of the component. However, some of the characteristics may not have a simple correspondence with the measureable material properties such as wear resistance, weldability and corrosion resistance which are not direct measurable parameters. In such case, functional properties are rated using an arbitrary scale. For example a corrosion rating of excellent, very good and good can be used as part of the properties of the material.

5.3.2 Manufacturing Requirements

The manufacturing of a material means provide a desired shape of material into a finished product. With reference to the specific manufacturing process several terminology such as castability, weldability, and machinability can be taken into consideration during the manufacturing of a product. Manufacturing process will always affect the material properties in service conditions and hence service life.

5.3.3 Reliability Requirements

The reliability requirements of the material are account for the fact that the product will perform within the intended function of the expected life without failure. Materials reliability is difficult to measure, because it is not only dependent upon the material's inherent properties, but also greatly affected by its production, processing history, environmental factors and transient overloading. Basically, new and nonstandard materials will tend to have 'lower reliability than established and standard materials.

5.3.4 Sustainable Requirement

The sustainable requirement plays an important role in determining the material performance requirements. This requirement is account for the action of hostile or other severe environment as the sensitivity of the environment will affect the materials' performance.

After analysis of the material performance requirements, it is recommended to classify the requirements into two main categories:

- Rigid or go-no-go requirement—this can be used for initial screening of the materials which will assist to get rid of unsuitable materials from the cluster or group. In fact this requirement should be considered firmly as certain materials are really out of choice.
- Soft or relative requirements—related to mechanical and physical properties of the materials, hence, it is understandable and negotiable.

5.4 Development of Different Solutions

There are two phases in developing different solutions for preliminary selection process, such as:

- Creativity phase
- Screening phase

5.4.1 Creativity Phase

Creativity phase deals with creating alternative solutions. Having specified the material requirements, the rest of the selection process involves searching the materials that would best meet those requirements. The starting point is the entire range of engineering materials. At this stage, creativity is essential to open up channels in different directions and not let traditional thinking but exploration of ideas. It should be done without much regard to their feasibility.

5.4.2 Screening Phase

Materials are sometimes chosen by trial and error or simply on the basis of what has been used before. While this approach frequently works, it does not always lead to optimization or innovation. In this phase, Ashby's material selection chart is used in order to choose the preliminary candidate materials which is known as screening.

The Ashby's approach is design-led approach. It starts by asking 'What is the function of the component in the design?', 'What objectives need to be optimized?', and 'What constraints must be satisfied?' For instance, a car body panel (function) needs to be as light as possible (objective) for a specified stiffness and cost (constraint). Other constraints on the design might be acceptable viz. resistance to mechanical impact and to contact with various environments. Figure 5.2 shows the easy visualisation of properties of engineering materials and ideal for a first 'rough cut' selection is known as Ashby's material selection chart.

The advantage of this approach is that it is systematic and unbiased in its focus on product objectives. The preliminary candidate material can be selected by using Ashby's material selection chart. Rigid requirement can be used for the initial screening of the materials to eliminate the unsuitable group of materials. From Fig. 5.3 is apparent that vertical axis strength refers to the young strength for mates and polymers, compressive strength for ceramics, tear strength for elastomers and tensile strength for composites with respects to the density of each material in the horizontal axis.

The common features of Ashby's material selection chart are (as can be seen in Fig. 5.3):

- Ashby charts are to be used only at the conceptual stage of selection of materials.
- There is a tremendous amount of information as well as power in these charts.
- Data are plotted on a log–log scale to include the enormous range of properties of different groups or classes of materials.
- Data for a given class of materials tend to cluster together on the charts.
- Data for a class are enclosed in a property envelope or property field. The envelope encloses all members of the class. Thick lines bound the envelopes.

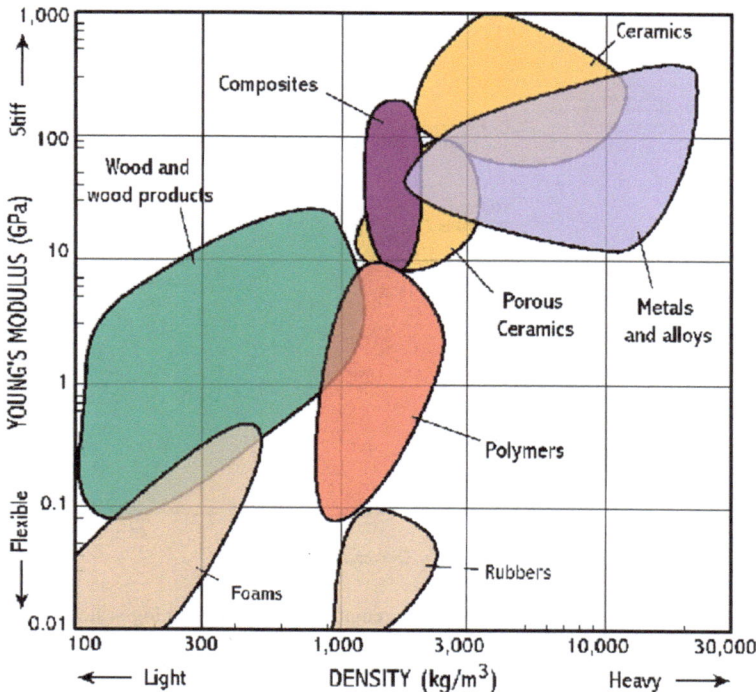

Fig. 5.2 Ashby's materials selection chart of elastic modulus versus density on log scale (Ashby 2005) (This material is reproduced with permission of John Wiley & Sons, Inc.)

- Within each envelope there are bubbles that signify the variation of properties of specific materials within a given class. A bubble encloses a typical range of properties for a single class of material. Light lines bound the bubbles.
- Overlap of envelopes sometimes occurs, but each class of materials has its own distinct field in the chart. Overlapping does not destroy the identity of a field.
- Each chart contains a set of design guidelines. The materials that are intersected by a given design guideline will perform equally well under the condition being considered. Materials above the line will show superior performance, while those lying below will exhibit poorer performances (and will be rejected during the screening stage).

Figure 5.4 shows the elastic modulus of polymers, metals, ceramics, and composites against density to show how the classes of materials group into common regions. Lines of constant slope are drawn on the diagram for different properties of material. For simple axial loading the relationship is $E/\rho = C$, for buckling of a slender column, $E^{1/2}/\rho = C$, and for the bending of a plate it is $E^{1/3}/\rho = C$. For instance, in determination of suitable materials for a column under compression, the slope would be $E^{1/2}/\rho = C$. In order to use this chart, it is

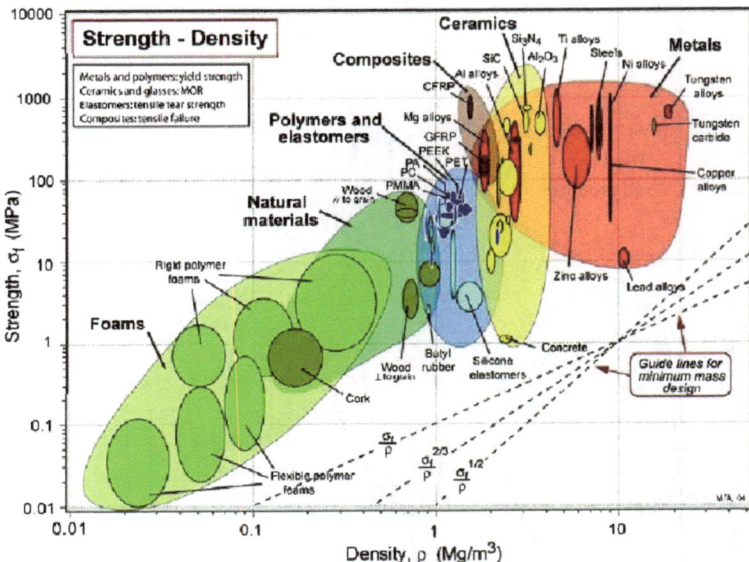

Fig. 5.3 Ashby's material selection chart of strength versus density on log scale. Yield strength for metals and polymers, compressive strength for ceramics, tear strength for elastomers and tensile strength for composites (Callister 1997, 4e) (This material is reproduced with permission of John Wiley & Sons, Inc.)

essential to start at the lower right-hand corner and move towards the upper left-hand corner. All the materials which lie on the line will perform equally well when loaded as a column in compression. Those materials which lie above the line are better, and those farthest from the line are the worst.

Based on Fig. 5.4, for simple axial loading, the relationship is E/ρ applies. For condition that could lead to buckling, $E^{1/2}/\rho$ applies, and for bending of a wide plate the relationship $E^{1/3}/\rho$ applies. Lines with these slopes are also can be seen in Fig. 5.4.

5.5   Quantitative Methods of Materials Selection

After narrow down of the field of possible candidate materials to those that do not violate any of the rigid requirements, one should start searching the material(s) that best meet the soft or relative requirements for best selection.

There are three quantitative methods for the evaluation of different solutions which are cost per unit property (CUP) method, weighted properties (WP) method, and digital logic (DL) method.

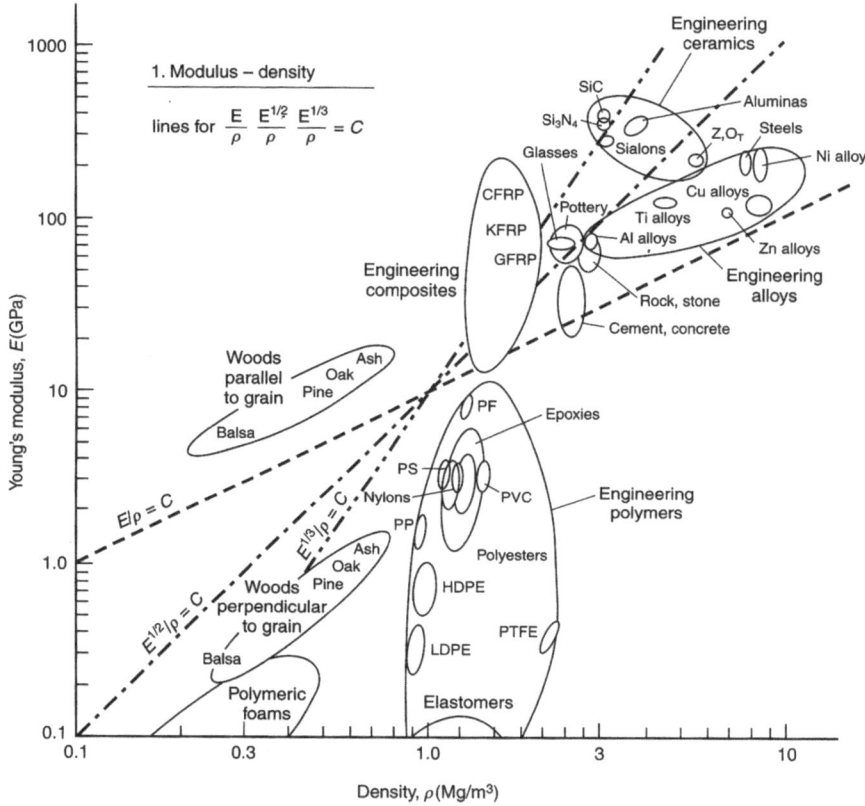

Fig. 5.4 Ashby materials selection chart: young's modulus versus density (From materials science and technology, vol. 5, 1989, pp. 517–525) (This material is reproduced with permission of Elsevier)

5.5.1 Cost Per Unit Property Method

This method can be used as a criterion for selecting the suitable material when one property stands out as the most critical service requirement.

It is the simplest cases of optimizing the selection of materials. Consider the case of a bar with the given length, L to support a tensile force, F. There are few assumptions such as:

- Cross-section of the material is solid cylindrical or bar
- Material is subjected to tension or compression
- Yield strength is the working stress or else it should be converted into working stress
- Cost of the material is related to the cost of material per unit mass.

Table 5.2 Formulas for estimating cost per unit property (Farag 2007)

Loading condition	Cost of unit strength	Cost of unit stiffness
Solid cylinder in tension/compression	$C\rho/S$	$C\rho/E$
Solid cylinder in bending	$C\rho/S^{2/3}$	$C\rho/E^{1/3}$
Solid cylinder in torsion	$C\rho/S^{2/3}$	$C\rho/G^{1/2}$
Solid cylindrical bar as slender column	–	$C\rho/E^{1/2}$
Solid rectangular in bending	$C\rho/S^{1/2}$	$C\rho/E^{1/3}$
Thin wall cylindrical pressure vessel	$C\rho/S$	–

where, S yield strength, E young's modulus, C cost per unit mass, ρ density and G shear modulus

The cross sectional is given by:

$$A = F/S \qquad (5.1)$$

where S is the working stress of the material, which is related to its yield strength by appropriate factor of safety. The cost of the bar is given by

$$\acute{c} = C\rho AL = (C \rho FL)/S \qquad (5.2)$$

where, C is cost of the material per unit mass and ρ is density of the material. In comparing different candidate materials, only the quantity ($C \rho/S$), which is the cost of unit strength, needs to be compared, as F and L are constant for all materials. The suitable material is the material which shows the lowest cost per unit strength.

If the applied load is alternating, it is more appropriate to use the fatigue strength of the material. Similarly, the creep strength should be used under loading conditions that cause creep. Equations similar to 2 and 3 can be used to compare materials on the basis of cost per unit stiffness when the important design criterion is deflection in the bar. In such cases, S is replaced by the elastic modulus of the material. Formulas for estimating cost per unit property are shown in Table 5.2.

The main limitation of the CUP method is that only one requirement (property) is considered as the most critical service requirement. In such case, the weighted property method may be useful tool for selecting suitable material.

5.5.2 Weighted Properties Method

Weighted properties (WP) method is useful to optimize materials selection when several properties are taken into consideration. Each material property is assigned a certain weight. *A weighted property value is obtained by multiplying the numerical value of the property by the weighting factor.* Sometimes scaling factors are used with numerical values. The normalized equation for the scale value of the properties is given in Eq. 5.3. After scaling the properties values of each materials are summed to give a comparative material performance index (P). Material with the highest P is considered as the best for the application.

In its simple form, the weighted properties method has the drawback of having to combine unlike units, which could yield irrational results. This is particularly true when different mechanical, physical and chemical properties with widely different numerical values are combined. The property with higher numerical value will have more influence than is warranted by its weighting factor. This drawback is overcome by introducing scaling factors. When evaluating a list of candidate materials, one property are scaled proportionally. Introducing a scaling factor facilitates the conversion of normal material property value to scaled dimensionless value. For a given property, the scaled value, B, for a given candidate material is equal to:

$$\text{Scaled property} = \frac{\text{Numerical value of property} \times 100}{\text{Maximum value in the list}} \qquad (5.3)$$

For properties like cost, corrosion or wear loss, weight gain in oxidation, etc., a lower value is more desirable. In such cases, the lowest value is rated as 100 it is calculated as:

$$\text{Scaled property} = \frac{\text{Minimum value in the list} \times 100}{\text{Numerical value of property}}.$$

Material performance index can be calculated using the following equation:

$$\text{Material performance index}, \gamma = \sum_{i=1}^{n} \beta_i \alpha_i \qquad (5.4)$$

where, β is the scaled property, α is the weighting factor and i is summed over all the n relevant properties. Though numerous material properties are considered but relative importance of each property is not defined clearly. Therefore, the weighting factors are become sensitive and as a result reduce the reliability of the material selection.

5.5.3 Digital Logic Approach

The digital logic (DL) method can be employed for the material selection with ranking. When numerous materials properties are specified and relative importance of each property is not clear determination of weighting factors can largely be sensitive which in turn will reduce the reliability of selection. DL approach can solve this problem to find the weighting factors. Only two properties are considered at a time. As a first step, the property requirements are determined based on the material selection chart. The total number of possible decisions is expressed as $N = n (n - 1)/2$, where n is the number of properties under consideration. The properties and the total number of decisions are shown in Table 5.3. The *weighting factor* for each property, which is indicative of the importance of one property as compared to others are obtained by dividing the numbers of positive decisions given to each property by the total number of decisions as shown in Table 5.4. After that calculate the scale value of each property followed by the materials performance index as explained in Sect. 5.5.2.

Table 5.3 Functional properties of candidate material and total number of positive decisions created by DL approach

Decision numbers

Properties	1	2	3	4	5	6	7	8	9	10
Tensile strength	0	0	0	1						
Yield strength	1				1	0	1			
Young's modulus		1			0			0	1	
Toughness			1			1		1		0
Density				0			0		0	1

Table 5.4 Weighting factor of the functional properties based on positive decision numbers

Properties of material	Positive decisions	Weighting factor (α)
Tensile strength	1	0.1
Yield strength	3	0.3
Young's modulus	2	0.2
Toughness	3	0.3
Density	1	0.1
Total	10	1.0

The cost of material should be emphasized when there are a large number of properties in the selection process which will lead to the further modification of the performance index of the material. Therefore, a figure of merit (FOM) for the material can then be calculated using the following formula:

$$M = \gamma/(C)(\rho) \tag{5.5}$$

where C = total cost of material per unit mass; ρ = density of material.

Alternatively, figure of merit can be calculated based on cost of unit strength (CUS) as we discussed under Sect. 5.5.1. The FOM then become:

$$M = \gamma/\acute{c} \tag{5.6}$$

where, $\acute{c}$ is the relative cost of the material and it is defined as the ratio of the price per unit mass of the material and low carbon steel.

5.6 Application of Digital Logic Method

5.6.1 Material Selection for Automotive Brake Disc: Case Study 1

Material used for brake systems should have stable and reliable frictional and wear properties under varying conditions of load, velocity, temperature, environmental factors and high durability. There are several factors to be considered when

Fig. 5.5 A solid cylindrical
shaft that experiences an
angle of twist in response to
the application of a twisting
moment M_t

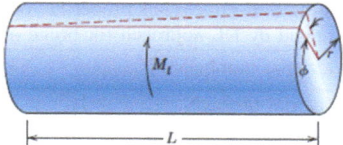

selecting a brake disc material. The most important consideration is the ability of
the brake disc material to withstand high friction and less abrasive wear. Another
requirement is to withstand the high temperature that evolved due to friction.
Weight, manufacturing process ability and cost are also important factors those are
need to be considered during the design phase. Figure 5.5 shows the advantage and
disadvantage of the digital logic method with comparing with the weight properties methods (Table 5.5).

Initial screening can be performed using Ashby's material selection chart (as
calculated in Fig. 5.4) and based on the properties, potential candidate materials
for automotive brake disc are selected as:

- Gray cast iron (GCI)
- Ti-alloy (Ti-6Al-4V)
- 7.5 wt% WC and 7.5 wt% TiC reinforced Ti-composite (TMC)
- 20 % SiC reinforced Al-composite (AMC 1)
- 20 % SiC reinforced Al–Cu alloy (AMC 2).

The properties and decision numbers are shown in Table 5.6 and corresponding
weighting factors for each property is shown in Table 5.7.

The numerical value of the each property of each candidate material is shown in
Table 5.8 whereas the calculated value of the scale properties of each material

Table 5.5 Advances and disadvantages of weighted property and digital logic methods

Method	Advantages	Disadvantages
Digital logic (DL) method	• Two properties are considered at a time • Systematic tool to determine weighting factor • All materials properties or performance goals are considered without keeping any shade of choice • Easier and faster method	• Might have some error as use own knowledge for making decision • The decision number for a property is given zero if it is not considered as important • Less accurate
Weighted properties (WP) method	• Determine weighting by giving certain weight with respective to the materials' properties impotence in service • Several properties can be considered at a time • Optimum scale value of the property can be obtained in preliminary step	• All units of properties are mixed and properties become sensitive • Provide inaccurate value

Table 5.6 Functional properties of candidate material and total number of positive decisions created by DL approach for brake disc material selection

Decision numbers, N

Material property	1	2	3	4	5	6	7	8	9	10
Compressive strength	0	0	0	1						
Friction coefficient	1				1	0	1			
Wear resistance		1			0			1	1	
Thermal capacity			1			1		0		0
Specific gravity				0			0		0	1

Table 5.7 Weighting factors based on positive decision numbers for brake disc

Property of material	Positive decisions	Weighting factor (α)
Compressive strength	1	0.1
Friction coefficient	3	0.3
Wear resistance	3	0.3
Thermal capacity	2	0.2
Specific gravity	1	0.1
Total	10	1.0

Table 5.8 Numerical value of the each property of brake disc materials

Material	Properties				
	1	2	3	4	5
	Compressive strength (MPa)	Friction coefficient (μ)	Wear rate ($\times 10^{-6}$ mm^3/N/m)	Specific heat, Cp (KJ/Kg. K)	Specific gravity (Mg/m^3)
GCI	1,293	0.41	2.36	0.46	7.2
Ti-6Al-4V	1,070	0.34	246.3	0.58	4.42
TMC	1,300	0.31	8.19	0.51	4.68
AMC 1	406	0.35	3.25	0.98	2.7
AMC 2	761	0.44	2.91	0.92	2.8

along with corresponding performance index (PI) is shown in Table 5.9. Finally, figure of merit (FOM) and ranking of the candidate materials for the application of automotive brake rotor system is shown in Table 5.10. From the Table, it can be seen that AMC 2 material showed highest PI followed by GCI. Finally, selection of optimum materials must be done. After narrowing down the field of possible candidate materials to those that do not violate any of the rigid requirements, one should start searching the material that best meet the soft requirements for optimum selection.

Table 5.9 Scaled value (SV) of properties of each material and corresponding performance index (PI)

| | Scaled value of each property and PI | | | | | |
	1	2	3	4	5	Performance index (γ)
GCI	99	93	100	47	38	81.0
Ti-6Al-4V	82	77	0.96	59	61	49.5
TMC	100	70	29	52	58	56.0
AMC 1	31	80	73	100	100	79.0
AMC 2	59	100	81	94	96	88.6

Table 5.10 Cost and figure of merit of candidate materials

Material	Relative cost	Performance index (γ)	Figure of merit	Rank
GCI	1	81.0	11.25	2
Ti-6Al-4V	20	49.5	0.56	5
TMC	20.5	56.0	0.58	4
AMC 1	2.7	79.0	10.84	3
AMC 2	2.6	88.6	12.17	1

5.6.2 Material Selection for Automotive Piston: Case Study 2

The aim of this case study is to select most appropriate material for the application of automotive piston emphasizing on the application of lightweight material. Initial candidate materials such as, gray cast iron, titanium alloy, aluminum alloy, composite materials such as 20 % SiC reinforced Al-composite, 20 % SiC reinforced Al–Cu alloy composite and WC-TiC reinforced Ti-composite are analyzed based on their performance requirements and alternative solutions are evaluated for the selection of most favorable material. A brief description of the potential candidate materials for automotive piston application is given below:

Cast iron: In order to improve performance, automotive engine is designed and operated by more power, higher pressure and higher piston speed. To achieve this grey cast iron (GCI) is beneficial and it is used as the main material for the piston, the graphite phase can act as an oil reservoir that supplies oil at dry starts or similar conditions of oil starvation. In view of the cost, manufacturing process and thermal stability, the GCI is actually a more specialized material for piston application particularly the material of choice for almost all automotive piston and piston ring applications.

Titanium alloys: Titanium alloys and their composites have the potential to reduce weight of the automotive component which is about 37–40 % less than a conventional cast iron with the same dimensions and offering high temperature strength and better resistance to corrosion. Titanium has a long record of successful performance in automotive application. This alloy is considered to be used for automotive piston due to higher strength, lower density and modulus. The use of titanium alloy is also increasing in many industrial and commercial applications due to its outstanding fatigue resistance and corrosion resistance in many environments especially in high strength applications.

Ni-Alloy: Presently nickel alloy contains less than 40 % Ni and its properties is much higher than the traditional Ni alloys which contained about 60 % of nickel. Recently introduced nickel has the excellent oxidation and creep properties which are very suitable for the use of industrial and automotive applications. It was specifically developed for maximum resistance to oxidation and resistance to oxide spallation when thermally cycled which is demanded for the automotive piston applications.

Al-Alloy (Al–Si eutectic alloy): Al–Si eutectic alloy occurs at 12.6 % Si and containing more than this percentage would be expected the primary silicon particles. Hypereutectic aluminum alloys are characterized by a eutectic matrix containing primary silicon crystals of various sizes and shapes. Among all the aluminum foundry alloys, Al–Si alloys constitute the major part of significant applications. It exhibits a highly desirable combination of characteristics such as castability, weldability, corrosion resistance and machinability. These alloys became popular as the chosen materials for automotive pistons because of their low thermal expansion coefficient, high strength to weight ratio and excellent wear resistance amongst the other properties.

Aluminium-metal matrix composite: The aluminium-metal matrix composite (AMC) materials having a lower density and higher thermal conductivity as compared to the traditionally used gray cast irons are expected to result in weight reduction of up to 60–65 % in automotive parts. The coefficient of thermal expansion decreases linearly with increase in SiC content in the composite. Higher speed movement from top dead center (TDC) to bottom dead center (BDC) of AMC piston can lower the friction coefficient value and cause a significant higher wear rate in the cylinder liner. The friction property of AMC piston is thus remarkable poorer than those of conventional cast iron material. Metal matrix composites provide the economic advantage over cast iron with respect to effective consumption of fuel and manufacturing cost. Metal matrix composites provide the economic advantage over cast iron with respect to effective consumption of fuel and manufacturing cost.

Ceramic Composite: Development of Ti-composite with 7.5 wt% WC and 7.5 wt% TiC reinforced is a promising means of achieving lightweight structural material combining high temperature strength with improved thermal shock resistance.

Tables 5.11 and 5.12 show the seven different common properties and relative cost of the engineering materials respectively, those are being considered for making an automotive piston. Table 5.13 shows the scaled value of the materials properties which is calculated using equation. Tables 5.13 and 5.14 shows the scaled value of the materials properties and positive decision respectively.

The weighting factor of each property is determined and shown in Table 5.15.

The figure of merit can be determined using previous formula and can be found the most appropriate material for the above automotive piston based on performance index and ranking. The performance index and ranking are shown in Table 5.16.

Table 5.11 Numerical values of the properties of candidate materials for automotive piston

Material	Properties						
	1	2	3	4	5	6	7
	Friction coefficient	Wear rate ($\times 10^{-6}$ mm^3/N/m)	Thermal capacity, (KJ/Kg-K)	Thermal conductivity (W/m-K)	Specific gravity (Mg/m^3)	Yield strength (MPa)	Tensile strength (MPa)
Gray cast iron	0.41	2.36	0.46	80	7.2	460	455
Ti-alloy	0.34	246.3	0.58	17.58	4.42	700	1,014
Ni-alloy	0.39	5.32	0.41	12.6	8.4	851	1,100
Al-alloy	0.36	2.89	0.714	190	2.73	480	510
Aluminium matrix composite 1	0.35	3.25	0.98	155	2.7	276	310
Aluminium matrix composite 2	0.44	2.91	0.92	180	2.8	425	485
Titanium matrix composite	0.31	8.19	0.51	17.85	4.68	700	1,029

Table 5.12 Relative cost of candidate materials

Material	Relative cost
Gray cast iron	1
Ti-alloy	20
Ni-alloy	17
Al-alloy	10
Aluminium matrix composite 1	2.8
Aluminium matrix composite 2	2.6
Titanium matrix composite	20.5

Table 5.13 Scaled value of the materials properties

Material Properties	Scaled values						
	1	2	3	4	5	6	7
GCI	93	100	47	42	42	54	39
Ti-alloy	77	0.96	59	9	61	82	92
Ni-alloy	89	44	42	7	32	100	100
Al-alloy	82	82	73	100	99	56	46
AMC 1	80	73	100	82	100	32	28
AMC 2	100	81	94	95	96	50	44
TMC	70	29	52	9	58	82	94

Table 5.14 Number of possible decision

Goals Property	Number of possible decision, [N = n(n − 1)/2]																				
	1	2	3	4	5	6	7	8	9	10	11	12	13	14	15	16	17	18	19	20	21
Friction coefficient	1	1	1	0	1	0															
Wear resistance	0						1	1	1	1	0										
Thermal capacity		0					0					1	0	1	0						
Thermal conductivity			0					0				0				1	0	0			
Specific gravity				1					0				1			0			1	0	
Yield strength					0					0				0			1		0		0
Tensile strength						1					1				1			1		1	1

Table 5.15 Weighting factor of each property

Property	Positive decisions	Weighting factor (α)
Friction coefficient	4	0.19
Wear resistance	4	0.19
Thermal capacity	2	0.095
Thermal conductivity	1	0.048
Specific gravity	3	0.14
Yield strength	1	0.048
Tensile strength	6	0.29
Total	21	1.00

Table 5.16 Performance index and ranking

Material	Relative cost	Performance index (γ)	Figure of merit	Rank
GCI	1	63.62	8.88	2
Ti-alloy	20	60.00	0.68	5
Ni-alloy	17	67.88	0.48	7
Al-alloy	10	72.44	2.65	4
AMC 1	2.8	66.16	8.75	3
AMC 2	2.6	76.72	10.54	1
TMC	20.5	63.50	0.66	6

The digital logic approach is implemented for the materials selection method with ranking based on performance index for the automotive piston. In DL method, performance index (PI) was calculated using the scaled property and weighting factor of each candidate materials. The DL method showed highest performance index for AMC 2 leading to the most appropriate material among seven proposed materials. The application of digital logic approach and knowledge-based system can be employed in many applications in industry including in the areas of design, manufacturing, trouble shooting and failure analysis besides the materials selection. However, modified digital logic method can also be implemented in these areas.

5.6.2.1 Design Example

A rectangular beam with a length of 0.5 m and width of 50 mm is subjected to a concentrated load of 10 kN which acts in the middle of the beam. The design guideline said that it should not suffer plastic deformation rather stiff upon the application of load. The yield strength (YS) and specific gravity of four different materials are shown in Table 5.17. Select the least expensive material of the beam. Assume the factor of safety of the beam is three. Write your comments on the selected material based on the cost per unit strength.

5.7 Optimum Material Selection

In this section, the performance index, γ, and the total cost, C_t of the candidate material are separately compared against currently used material. Since the purpose is to improve performance, acceptable candidates must perform at a higher level than the currently used material. If cost is not the objective, the candidate with the highest performance index can be selected.

The percentage increase in performance ($\Delta\gamma$ %) and the corresponding percentage increase in cost (ΔC_t %) for both the candidate materials and the currently used material is calculated using following Eqs. (5.5) and (5.6) respectively and

Table 5.17 Properties of materials

Materials	MOE (GPa)	Specific gravity	Relative cost	Cost of unit stiffness $[(\hat{c})(\rho)/E^{1/3}]$
Steel AISI 1020	207	7.86	1	–
Steel AISI 4140	390	7.86	1.5	–
Al 6061 T6	70	2.7	6	–
Graphite fibre reinforced composite	25	2.11	9	–

Table 5.18 The percentage increase in performance ($\Delta\gamma$ %) and the corresponding percentage increase in cost (ΔC_t %) for both the candidate materials

Materials	Performance index (γ)*	Total cost (£/ton) (C_t)	$\Delta\gamma$ %	ΔC_t %
GCI	81.0	830	9.4	167.5
MMC2	88.6	2,220		

Note * these values are taken from Table 5.10

values are summarized in tabular form. Table 5.18 shows percentage increase in performance ($\Delta\gamma$ %) and the corresponding percentage increase in cost (ΔC_t %) for a new (such as MMC2) and current (GCI) materials.

$$\Delta\gamma\% = \frac{100(\gamma_n - \gamma_0)}{\gamma_0} \tag{5.7}$$

$$\Delta C_t\% = \frac{100(C_{tn} - C_{t0})}{C_{t0}} \tag{5.8}$$

where,
γ_n and γ_0 Performance indices of the new and original materials
C_{tn} and C_{t0} Cost of the new and original materials, respectively.

5.8 Material Selection for a Cylindrical Shaft: Case Study 3

Shafts are the important elements of machines because they are the element that support rotating parts like gears and pulleys and in turn are themselves supported by bearings resting in the rigid machines housings. The function of a shaft is to transmit power from one rotating member to another supported by it or connected to it. Thus, they are subjected to torque due to power transmission and bending moment due to reactions on the members that are supported by them. Shaft can be classified into straight shaft, cranked, flexible or articulated shaft with always made to have circular cross-section and it could be either solid or hollow. Straight shafts are commonest to be used for power transmission. Such shafts are commonly designed as stepped cylindrical bars, that is, they have various diameters along their length, although constant diameter shafts would be easy to produce.

The shafts are always subjected to fatigue load hence they must be calculated for fatigue strength under combined bending and torsion loading. However, the initial estimate of diameter is obtained from the torque that is transmitted by the shaft. The bending moment variation along the length of the shaft is established after fixing some structural features like distance between supporting bearings and distance between points of application of forces and bearings.

In this case study, the component or structural element that has been chosen to be discussed is a solid cylindrical shaft that is subjected to a torsional stress. Strength of the shaft will be considered in detail, and criteria will be developed for maximizing strength with respect to both minimum material mass and minimum cost. Other parameters and properties such as stiffness that may be important in this selection process are also discussed briefly.

For this problem, a criterion for selection of light and strong materials of the shaft will be established. It will be assumed that the twisting moment and length of the shaft are specified, whereas the radius (or cross-sectional area) may be varied. We develop an expression for the mass of material required in terms of twisting moment, shaft length, and density and strength of the material. Using this expression, it will be possible to evaluate the performance—that is, maximize the strength of this torsionally stressed shaft with respect to mass and, in addition, relative to material cost.

Consider the cylindrical shaft of length L and radius r, as shown in Fig. 5.5.

The application of twisting moment (or torque), M_t produces an angle of twist ϕ. Shear stress τ at radius r is defined by the equation

$$\tau = \frac{M_t r}{J} \tag{5.9}$$

Here, J is the polar moment of inertia, which for a solid cylinder is;

$$J = \frac{\pi r^4}{2} \tag{5.10}$$

$$\tau = \frac{2M_t}{\pi r^3} \tag{5.11}$$

A safe design calls for the shaft to be able to sustain some twisting moment without fracture. In order to establish a materials selection criterion for a light and strong material, we replace the shear stress in Eq. 5.10 with the shear strength of the material τ_f divided by a factor of safety N, as

$$\frac{\tau_f}{N} = \frac{2M_t}{\pi r^3} \tag{5.12}$$

It is now necessary to take into consideration material mass. The mass m of any given quantity of material is just the product of its density (ρ) and volume.

Since the volume of a cylinder is just $\pi r^2 L$, then

$$m = \pi r^2 L \rho \tag{5.13}$$

Or, the radius of the shaft in terms of its mass is just

$$r = \sqrt{\frac{m}{\pi L \rho}} \tag{5.14}$$

Substitute r into Eq. 5.12 leads to

$$\frac{\tau_f}{N} = \frac{2M_t}{\pi \left(\sqrt{\frac{m}{\pi L \rho}} \right)^3} \tag{5.15}$$

$$= 2M_t \sqrt{\frac{\pi L^3 \rho^3}{m^3}}$$

Solving this expression for the mass m yields

$$m = (2NM_t)^{2/3} \left(\pi^{1/3} L \right) \left(\frac{\rho}{\tau_f^{2/3}} \right) \tag{5.16}$$

The parameters on the right-hand side of this equation are grouped into three sets of parentheses. Those contained within the first set (i.e., N and M_t) relate to the safe functioning of the shaft. Within the second parentheses is L, a geometric parameter. And, finally, the material properties of density and strength are contained within the last set. The upshot of Eq. 5.16 is that the best materials to be used for a light shaft which can safely sustain a specified twisting moment are those having low ratios. In terms of material suitability, it is sometimes preferable to work with what is termed a performance index, P, which is just the reciprocal of this ratio; that is

$$P = \frac{\tau_f^{2/3}}{\rho}. \tag{5.17}$$

In this context we want to utilize a material having a large performance index. At this point it becomes necessary to examine the performance indices of a variety of potential materials. This procedure is expedited by the utilization of what are termed materials selection charts.

These are plots of the values of one material property versus those of another property. Both axes are scaled logarithmically and usually span about five orders of magnitude, so as to include the properties of virtually all materials. For example, the chart of interest is logarithm of strength versus logarithm of density, which is shown in Fig. 5.6. It may be noted on this plot that materials of a particular type (e.g., woods, engineering polymers, etc.) cluster together and are enclosed within an envelope delineated with a bold line. Subclasses within these clusters are enclosed using finer lines. Now, taking the logarithm of both sides of Eq. 5.17 and rearranging yields,

$$\log \tau_f = \frac{3}{2}\log \rho + \frac{3}{2}\log P \qquad (5.18)$$

5.8.1 Screening of Candidate Materials

This expression tells us that a plot of $\log \tau_f$ versus $\log \rho$ will yield a family of straight and parallel lines all having a slope of each line in the family corresponds to a different performance index (P). These lines are termed design guidelines, and four have been included in Fig. 5.6 for P values of 3, 10, 30, and 100 $(MPa)^{2/3}m^3/Mg$. Relative to materials selection, all materials that lie on one of these lines will perform equally well in terms of strength-per-mass basis; materials whose positions lie above a particular line will have higher performance indices, while those lying below will exhibit poorer performances. For example, a material on the P = 30 line will yield the same strength with one-third the mass as another material that lies along the P = 10 line.

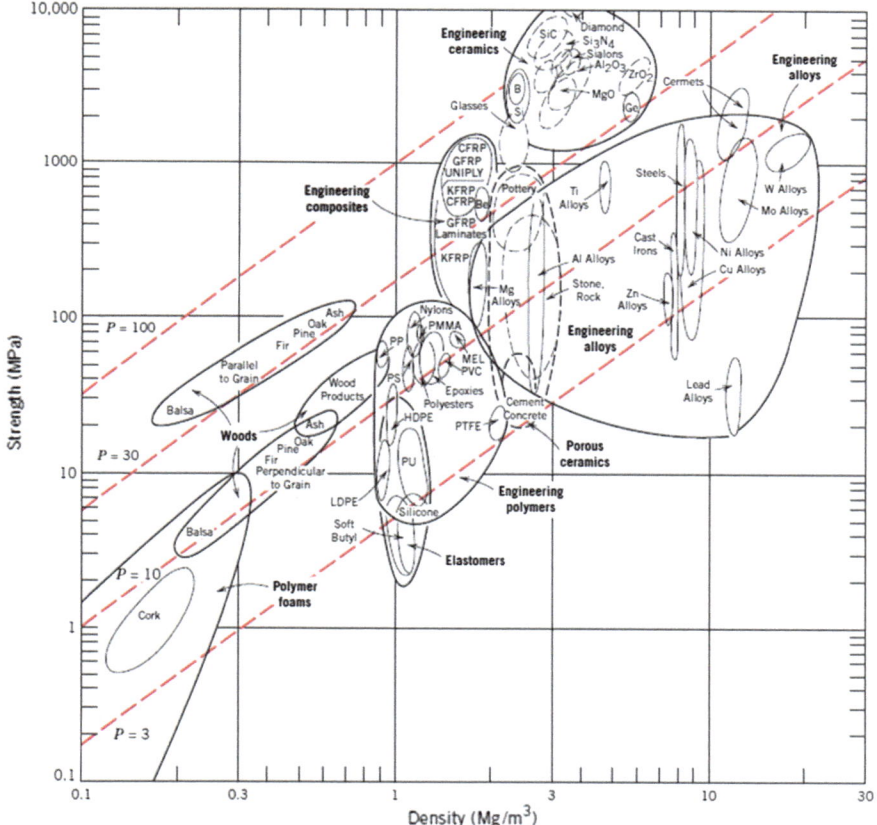

Fig. 5.6 Strength versus density materials selection chart (Callister 1997, 4th edition). (This material is reproduced with permission of John Wiley & Sons, Inc.)

Table 5.19 Density (ρ), strength (τ_f), and performance index (P) for five engineering materials

Materials	ρ (Mg/m^3)	τ (MPa)	$P = \tau_f^{2/3}/\rho$
Carbon fiber reinforced composite	1.5	1,140	72.8
Glass fiber-reinforced composite	2.0	1,060	51.2
Aluminum alloy (2024-T6)	2.8	300	16.0
Titanium alloy (Ti-6Al-4V)	4.4	525	14.8
AISI 4340 steel[*]	**7.8**	**780**	**10.9**

[*] This materials P complies with design guideline

The selection process now involves choosing one of these lines, a "selection line" that includes some subset of these materials; within the range of 10 (MPa)$^{2/3}$ m^3/Mg, which is represented in Fig. 5.6. Materials lying along this line or above it are in the "search region" of the diagram and are possible candidates for this rotating shaft. These include wood products, some plastics, a number of engineering alloys, the engineering composites, and glasses and engineering ceramics. On the basis of fracture toughness considerations, the engineering ceramics and glasses are ruled out as possibilities. All wood products, all engineering polymers, other engineering alloys (viz. Mg and some Al alloys), as well as some engineering composites can be eliminated due to their lower strength and finally the potential candidate materials are; steels, titanium alloys, high-strength aluminium alloys, and the engineering composites.

Table 5.19 presents the density, strength, and strength performance index for three engineering alloys and two engineering composites, which were deemed acceptable candidates from the analysis using the materials selection chart in order to evaluate and compare the strength performance behavior of specific materials. The five materials in Table 5.19 are ranked according to strength performance index, from highest to lowest: carbon fiber-reinforced and glass fiber-reinforced composites, followed by aluminum, titanium, and AISI 4340 steel alloys. However, from the Table 5.19, it is clear that AISI 4340 steel meets the design specification with the performance index value of 10.9 and hence, this material can be consider as an appropriate material for the design of shaft which is subjected to twisting moment. However, further evaluation is important after considering the factor of cost.

One way to determine materials cost is by taking the product of the price (on a per-unit mass basis) and the required mass of material. Cost considerations for these five remaining candidate materials—steel, aluminum, and titanium alloys, and two engineering composites—are presented in Table 5.20. In the first column of the table, the value is $\tau_f^{2/3}/\rho$. The next column lists the approximate relative cost, denoted as $\hat{c}$, this parameter is simply the per-unit mass cost of material divided by the per-unit mass cost for low-carbon steel, one of the common engineering materials. The underlying rationale for using $\hat{c}$ is that while the price of a specific material will vary over time, the price ratio between that material and another will, most likely, change more slowly. The product of $\rho/\tau_f^{2/3}$ and $\hat{c}$ is basically refer to the cost per unit strength of the materials which we explained earlier in Sect. 5.5.1.

The most economical or least cost material is AISI 4340 steel followed by the glass fiber-reinforced composite, AA 2024-T6 aluminum, the carbon

Table 5.20 Tabulation of the $\rho/\tau_f^{2/3}$ R ratio, relative cost ($\hat{c}$), and the product of $\rho/\tau_f^{2/3}$ and $\hat{c}$ for five engineering materials

Materials	$\rho/\tau_f^{2/3}$	$\hat{c}$	$\hat{c}\,(\rho/\tau_f^{2/3})*$
AISI 4340 Steel	**9.2**	**5**	**46**
Glass fiber-reinforced composite	1.9	40	76
Aluminum alloy (2024-T6)	6.2	15	93
Carbon fiber-reinforced composite	1.4	80	112
Titanium alloy (Ti-6Al-4V)	6.8	110	748

This product refers to the least cost material after being considered the cost factor

fiber-reinforced composite, and the titanium alloy. Thus, when the issue of cost is considered, there is a significant alteration within the ranking scheme. For example, the carbon fiber-reinforced composite is relatively expensive, it is significantly less desirable; or, in other words, the higher cost of this material may not outweigh the enhanced strength it provides.

5.9 Summary

The fundamental criteria and concept of materials selection process have been discussed. Comprehensive quantitative materials selection methods such as cost per unit property (CUP), weighted property (WP) and digital logic (DL) method have been explained with suitable examples. The advantages and disadvantages of WP and DL methods are also highlighted in this chapter. A systematic case study on the material selection of automotive brake rotor system has been presented for ease understanding and implementation of the materials selection method in any engineering applications as well. An expression for the strength performance index is derived for a torsionally stressed cylindrical shaft; then, using the appropriate materials selection chart, a preliminary candidate search was conducted. From the results of this search, several candidate engineering materials were ranked on both strength-per-unit mass and cost per unit strength. Other factors that are relevant to the decision-making process were also discussed.

5.10 Tutorial Questions

5.1   What do you mean by relative cost of materials?

5.2   Write down the advantages and disadvantages of digital logic (DL) method over weighted properties (WP) method for the evaluation of different solution in selection of materials.

5.3   Explain the important features of Ashby's materials selection chart.

5.4   Write down the different stages of materials selection process. For the analysis of material performance requirements the term 'performance index' can be used. In generally, the material performance requirements can

Table 5.21 Candidate materials for automotive brake pad application

Materials	Specific gravity (kg/m^3)	Yield strength (MPa)	Relative cost
Alumina whiskers	3.8	300,000	3.5
CFRP	2.0	544	4.5
4340 Steel	7.8	296	1
Al-alloy	4.5	420	6

be divided into five broad categories. List down all of them and explain only two (2).

5.5 In the first stages of material selection process development, one should find out the fundamental criterion and concepts of material selection by answering the following questions:

WHY
WHEN
WHO and
HOW

In line with these, discuss the fundamental criterion and concepts of material selection process. For the analysis of material performance requirements the term 'performance index' can be used. List down all material performance requirements those are considered for the materials selection purposes and explain only two (2) of them.

5.6 A bicycle manufacturer in Malaysia is considering the possibility of using fibre-reinforced polymer composite (FRPC) in making bicycle frames. Write down the main functional requirements and corresponding material properties suitable for the above application.

5.7 Table 5.21 shows the candidate materials for automotive brake pad application. Using Cost per Unit Property (CUP) method, find the least expensive material of the automotive brake pad.

5.8 Explain the weighted properties (WP) method for the selection of materials when several properties are taken into consideration.

5.9 What are the main functional requirements and corresponding material properties need to be considered for the following products: (a) automotive brake rotor, (b) automotive piston, (c) automotive cylinder liner, and (d) an aircraft wing structure?

5.10 Explain the design consideration of materials selection for the following products: (a) tennis racket, (b) automotive bumper and (c) bicycle frame.

5.11 The following Table 5.22 for four materials along with their properties are being considered for making a structural component.

(a) Calculate the scaled values of the properties of material.
(b) Determine the weighting factors for the above properties.
(c) Using the method of weighted properties, find the best candidate material for the above structural application.

Table 5.22 Structural component materials

Properties	Materials			
	Al-alloy	Steel AISI 1015	Graphite fibre reinforced composite	Kevlar fibre reinforced composite
Yield strength (MPa)	248	329	680	425
Modulus of elasticity (GPa)	70	207	22	27
Weldability index (5 = excellent, 4 = very good, 3 = good)	3	5	4	4
Specific gravity	2.8	7.8	2.0	2.3

Table 5.23 Common engineering materials along with their properties

Properties	Materials			
	Al 2014-T6 alloy	Steel AISI 1015	Graphite fibre reinforced composite	Natural fibre reinforced composite
Yield strength (MPa)	248	329	680	425
Modulus of elasticity (GPa)	70	207	22	27
Weldability index (5 = excellent, 4 = very good, 3 = good)	3	5	4	4
Specific gravity	2.8	7.8	2.0	2.3

5.12 The Table 5.23   shows the common engineering materials along with their properties   and are being considered for making automotive bottom structure (such as frame and beam).

Using the above information in the table,

(a) Calculate the scaled values of the properties of material.
(b) Determine the weighting factors for the above properties.
(c) Using the weighted factors for the different properties, what would be the best material?

5.13 What are the main design requirements for an automotive brake disc/rotor? Select the optimum material out of the following five candidate materials for use in fabricating the brake rotor: Table 5.24 shows the properties of candidate materials for brake disc.

Compare and justify your optimum selection of material using knowledge-based expert system.

Table 5.24 Properties of candidate materials for brake disc

Materials	Properties				
	Compressive strength (MPa)	Friction coefficient (μ)	Wear rate ($\times 10^{-6}$ mm^3/ N/m)	Specific heat, Cp (KJ/Kg. K)	Specific gravity (Mg/m^3)
Gray Cast iron	1,293	0.41	2.36	0.46	7.2
Ti-6Al-4V	1,070	0.34	246.3	0.58	4.42
Ti composite	1,300	0.31	8.19	0.51	4.68
AMC 1 (20 %SiC$_p$-Al)	406	0.35	3.25	0.98	2.7
AMC 2 (20 % SiC$_p$-Al-Cu matrix composite)	761	0.44	2.91	0.92	2.8

5.14 The following Table 5.25 for five materials along with their properties are being considered for making a drinking container.

(a) Calculate the scaled values of the properties of material.
(b) Determine the weighting factor of decisions for the above properties.
(c) Using the digital logic method, find the optimum candidate material for the above drinking container.

5.15 Figure 5.7 shows a typical plot of Strength versus Density which is known as Ashby's Materials selection chart.
Using the above chart answer the following

(a) How do you represent the ovals in the chart? Give examples.
(b) Why do you use the dotted lines in the chart?
(c) Why do you use the Ashby's chart as a whole?

Table 5.25 Five materials along with their properties are being considered for making a drinking container

Properties	Materials				
	Aluminium (3104)	Steel (440A)	Glass	Polyethylene, (PE)	Polyethanol, PET
Tensile strength (MPa)	180	186	82.74	8	50
Thermal expansion ($10^{-6}k^{-1}$)	23.1	17.3	8.5	84.5	59.4
Production energy (MJ/kg)	200	23	14	80	84
Specific gravity	2.7	7.96	2.53	0.95	1.5

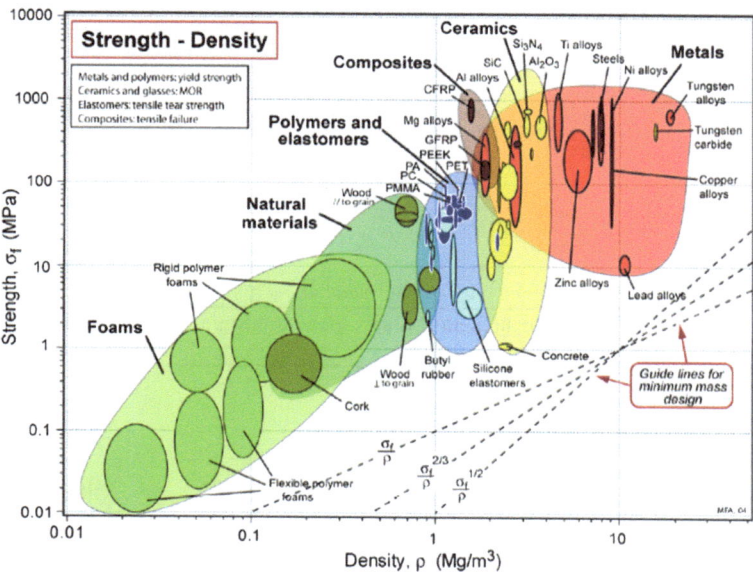

Fig. 5.7 Plot of strength versus density

Table 5.26 Soldering materials and their properties

Solder alloy	Melting point, °C (liquidus)	Density, g/cm^3	Electrical resistivity, $\mu\Omega \cdot m$	Thermal conductivity, W/m·K	Tensile strength at break, kgf/cm^2
Sn96.5Ag3.5 (alloy #121)	221	7.37	0.123	55	580
Bi58Sn42 (alloy #281)	138	8.56	0.383	19	565
Sn63Pb37 (alloy #106)	183	8.40	0.145	50	525
Bi55.5Pb44.5 (alloy #255)	124	10.44	0.431	4	450
Sn37.5Pb37.5In25 (alloy #5)	181	8.42	0.221	23	370
Sn95Sb05 (alloy #133)	240	7.25	0.145	28	415

5.16 The Table 5.26 five materials along with their five properties are being considered for soldering in chip fabrication process.

    (a) Calculate the scaled value of each property of material.

    (b) Determine the weighting factor of decisions for the above properties.

    (c) Using the digital logic method, find the optimum candidate material for the above soldering in chip fabrication process.

Fig. 5.8 Cross-sectional view of a solid cantilever beam

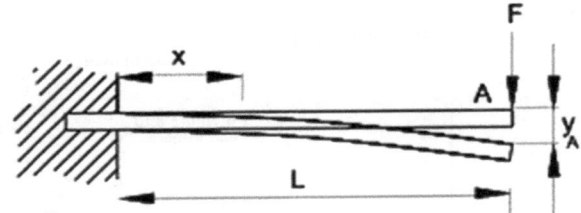

Table 5.27 Properties of common engineering materials

Material	Shear strength (MPa)	Density (Mg/m³)	Relative cost of material
Graphite fibre reinforced composite	1,060	2.0	6.5
Polymer matrix composite	1,020	1.8	6.0
Al-alloy (2026-T6)	300	2.8	10
Ti-alloy	480	4.5	20
Alumina whiskers	680	3.1	40
4,340 steel (Q&T)	780	7.8	2

5.17 Figure 5.8 shows a cross-sectional view of a solid cantilever beam. For the protocol of the material selection, derive a mass-strength performance index of this cantilever beam which is subjected to deflection.

Table 5.27 shows the properties of several engineering materials. On the basis of the performance index (as you derived in Q5.17) and using Table 5.27, rank the materials and write your comments.

References

Ashby MF (1989) On the engineering properties of materials. Acta Metall 37:1273

Ashby MF (2005) Materials selection in mechanical design, 3rd edn, Butterworth-Heinemann, London

Callister WD (1997) Materials science and engineering: an introduction, 4e, Wiley, London

Farag MM (2007) Selection of materials and manufacturing processes for engineering design. Prentice Hall, London

Ghandhi MV, Thompson BS (1992) Smart materials and structures. Chapman and Hall, UK, pp 1–4

Lennart Y, Edwards KL (2003) Design, materials selection and marketing of successful products. Mater Des 24:519–529

Chapter 6
Knowledge Based System in Materials Selection

Abstract This chapter reviews the knowledge based expert systems (KBS) in material selection with emphasize on rule-based expert system for the selection of material. This chapter begins with expert system component fact and expert system architecture followed by benefits of knowledge based expert system. Several case studies on the application of KBS for optimum selection of materials have been addressed.

Keywords Expert system architecture · Rule-based system in materials selection · Analytical hierarchy method · Case study

Learning Outcomes

After learning this chapter student should be able to do the following:

> Identify the importance of knowledge based system
> Explain the expert system architecture and rule based system
> Apply the knowledge based system for optimum selection of materials

6.1 Introduction

A knowledge based system (KBS) or expert system is a problem solving and decision making system based on knowledge of its task and logical rules or procedures for using knowledge. A knowledge based system that represents expertise is referred to as expert system. Knowledge-based system is defined as (Gonzalez and Dankel 1993):

> A computerized system that uses knowledge about some domain to arrive at a solution to a problem from that domain. This solution is essentially the same as that concluded by a person knowledgeable about the domain of the problem when confronted with the same problem.

Many expert systems are not complex or difficult to build. In a very simple case, consider a tree diagram on paper describing how to solve a problem. By making a selection at each branch point, the tree diagram can help someone make a decision.

M. A. Maleque and M. S. Salit, *Materials Selection and Design*,
SpringerBriefs in Materials, DOI: 10.1007/978-981-4560-38-2_6,
© The Author(s) 2013

In a sense, it is a very simple expert system. This type of tree structured logic can easily be converted into a computerized system that is easier to use, faster and automated. More elaborate systems may include confidence factors allowing several possible solutions to be selected with different degrees of confidence. Expert systems can explain why data is needed and how conclusions are reached. A system may be highly interactive (directly asking the user questions) or embedded where all input comes from another program. The range of problems that can be handled by expert systems is vast. Expert systems software can be developed for any problem that involves a selection from among a definable group of choices where the decision is based on logical steps.

Any area where a person or group has special expertise needed by others is a possible area for an expert system. Expert systems or knowledge based system can help to automate anything from complex regulations for assisting users in selecting from among a group of products, or diagnosing equipment problems. Traditionally expert system development is a major expense both in time and money. The key to implement expert systems widely, effectively and at a low cost is to have easy-to-use expert system development tools readily available to the experts. As more power is needed for certain applications, higher level tools can be used with advanced features to give complete control over the inference engine, modularization of the knowledge base, flow of execution, the user interface and integration with other programs.

Knowledge based systems are artificial intelligent tools working in a narrow domain to provide intelligent decisions with justification. Knowledge is acquired and represented using various knowledge representation techniques, rules, frames and scripts. The basic advantages offered by such system are documentation of knowledge. Knowledge is the most valuable of all corporate resources that must be captured, stored, re-used and continuously improved, in much the same way as database systems were important in the previous decade.

Flexible, extensible, and yet efficient knowledge base systems are needed to capture the increasing demand for knowledge-based applications which will become a significant market in the next decade. Knowledge can be expressed in many static and dynamic forms; the most prominent being domain objects, their relationships, and their rules of evolution and transformation. It is important to express and seamlessly use all types of knowledge in a single knowledge base system.

6.2 Expert System Components Facts

6.2.1 Rule-Based Reasoning

The most common form of expert system structure is rule-based expert system. The features of rule based system are:

- It provides user defined interface
- User friendly

- Consider as an "Intelligent" system
- Express knowledge to present the information
- Use knowledge based on preference.

6.2.2 Database

- Contains some of the data of the interest to the system
- Connected to on-line user profile or public database
- Appropriate for human user.

6.2.3 Inference Engine

- General problem-solving knowledge methods
- Interpret, analyze and process the rules
- Determines Schedule which rule to look at next
- Can search certain portion of a rule-based system
- Provide exploratory information.

6.3 Architecture and Elements in Expert System

Knowledge-based systems are finding many applications in industry including in the areas of design, manufacturing and materials selection. Model development for the prediction of output of the machining is also among the areas that lend themselves to the application of expert systems (Krishna et al. 2009). The knowledge and the logic are obtained from the experience of a specialist in the area. They usually consist of a set of knowledge sources, a knowledge base including a meta-knowledge base and a reasoning system for an inference mechanism. There are several forms of expert systems that have been classified according to the methodology used including: rule-based systems, case-based reasoning systems and fuzzy expert systems. Typical expert system architecture, components and human interface is shown in Fig. 6.1. From the Fig. 6.1, it can be presumed that working storage techniques (databases) and knowledge bases provide the basis of an expert system whereas the inference engine is accessible to the user through a user interface. Figure 6.2 shows an expert network elements which is used in conjunction with an expert system. The expert system provides consistency checks and also does inventory and economic analysis. The expert system has two main modules such as, material analyzer module and inventory check module. However, there are three other supporting modules for the expert systems which include update module, explanation module and help module.

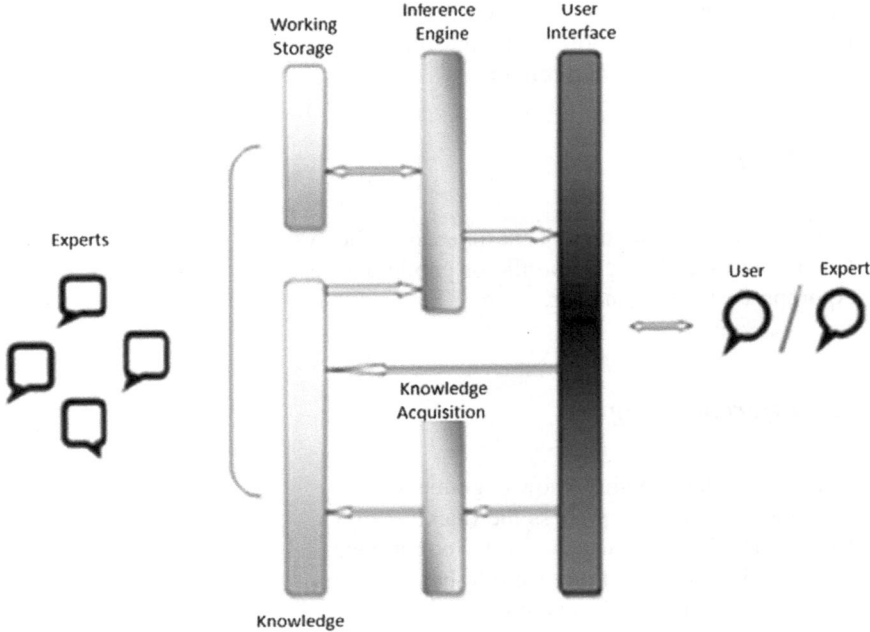

Fig. 6.1 Conventional architecture of the expert system, components and human interface (Durkin 1996)

6.4 Benefits of Knowledge Based System

The following benefits can be obtained from KBS:

- Less energy and less time consuming
- Obtain immediate results
- Can optimize the results easily
- Repeatedly can use or apply KBS for any design or application
- KBS is an efficient and effective process for wide range of applications
- Provides recommendations on the optimum solution of the selection.

However, for the rule-based system, at least two conditions are to be considered such as functional requirement definition and availability of the materials' properties in the database. Interface engine and user interface allow the user to input relevant variables.

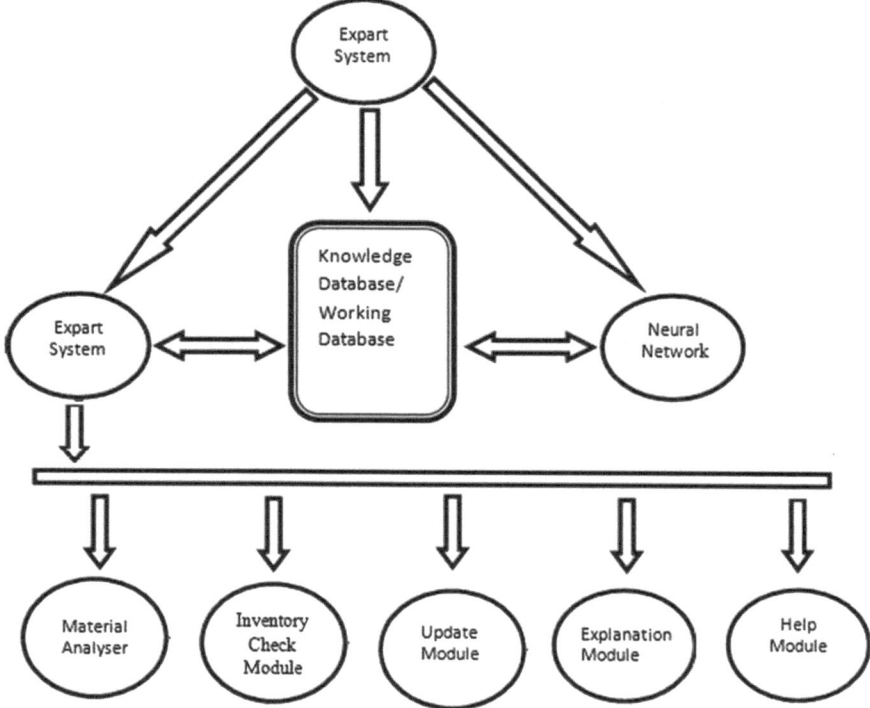

Fig. 6.2 Elements of the expert network

6.5 KBS for Optimum Selection of Materials: Case Study 1

This study describes the use knowledge based system to select an optimum material for engine components such as piston, connecting material and piston ring from ceramic matrix composites. This proposed knowledge based system material selection consists of a user interface, knowledge acquisition, inference engine, knowledge base and database as shown in Fig. 6.3. The knowledge based system tool-kit used in this study is Kappa-PC developed by Intellicorp Inc.

Fig. 6.3 Structure of the knowledge-based system (Sapuan et al. 2002) (reprinted with permission of Elsevier Science)

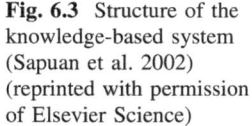 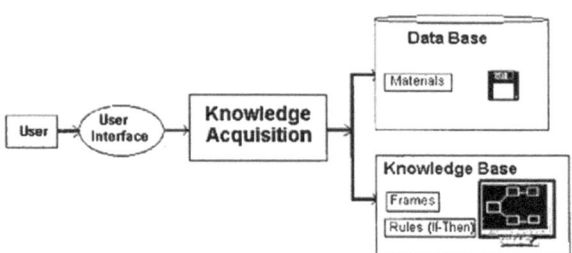

6.5.1 User Interface

A user friendly interface is designed in such a way that when an image in the interface is clicked, it invokes the necessary function and starts running the inference process. Figure 6.4 gives a user interface for the materials selection process.

6.5.2 Inference Engine

The inference engine works based on rules and finds the solution by reasoning method. It looks through the knowledge base and finds the solution with respect to the rule constraints. The reasoning continues until the suitable results are obtained based on the constraints pre-defined in the rules.

6.5.3 Database

The system database consists of four separate classes of ceramic matrix composites, namely database for:

- Fiber reinforcement composite
- Whisker reinforcement composite
- Particle reinforcement composite
- Platelet reinforcement composite.

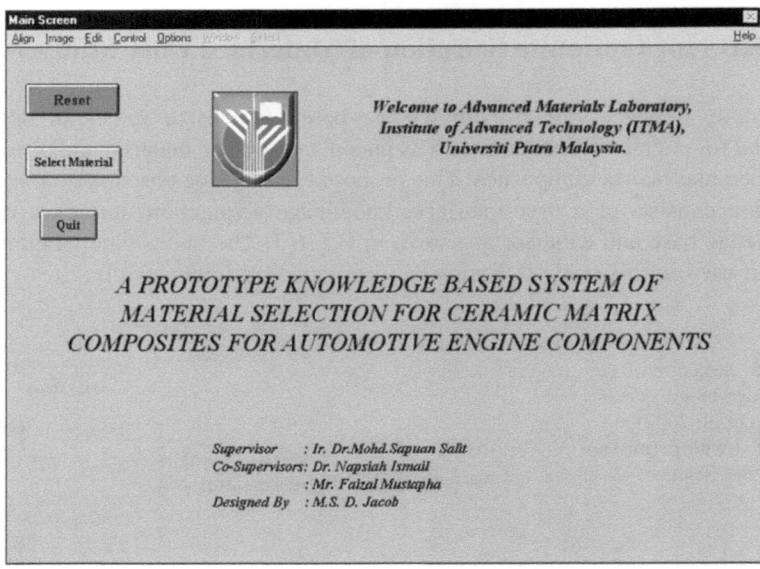

Fig. 6.4 User interface (Sapuan et al. 2002) (reprinted with permission of Elsevier Science)

6.5.4 Knowledge Base

The knowledge base consists of rules and techniques for representing knowledge in the materials selection system. Information related to ceramic matrix composites and the types of reinforcements are represented as objects, and their properties such as manufacturing techniques, mechanical and physical properties are stored in the knowledge base. The relationship between the objects is linked together by a way of hierarchy. Figure 6.5 shows the object hierarchy of the proposed system.

6.5.5 Material Selection Process

The rule based system is the major important tool in the materials selection system. For each engine component studied, several rules are created. These are chained by using forward chaining technique. The constraints or limiting values are selected from the product design specification (PDS). The simple rule can be presented as:

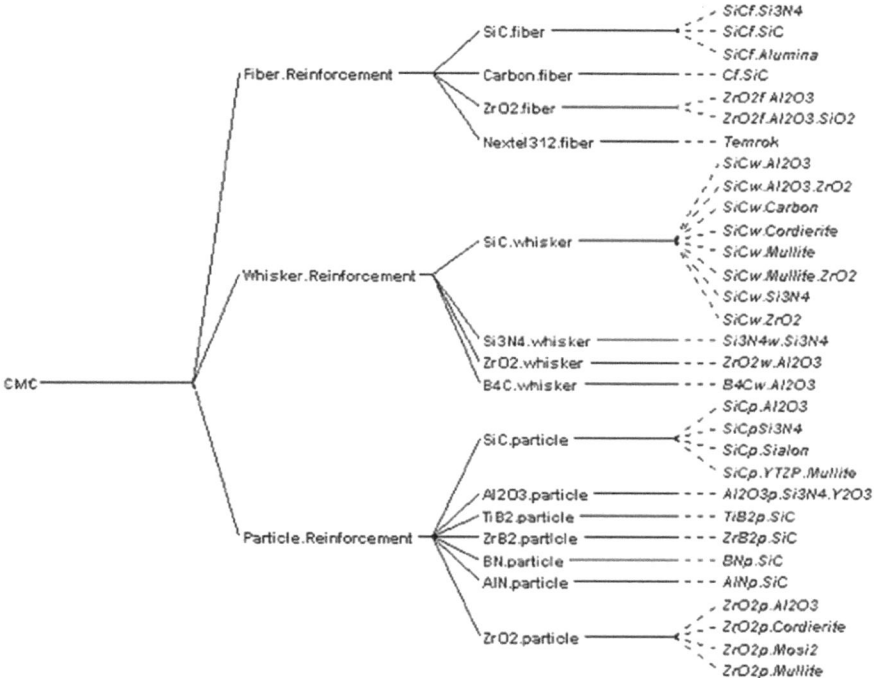

Fig. 6.5 Object hierarchy of the proposed system (Sapuan et al. 2002) (reprinted with permission of Elsevier Science)

If
(condition)
Then
(conclusion)

If the condition of a rule is satisfied, then the conclusion of the rule is set as the result. If the constraint values satisfy more than one material, then the related materials are selected as results and ranked according to the properties and stored in the transcript images. For ranking the materials, the properties such as tensile strength, fracture toughness and Young's modulus are considered. Figure 6.6 shows a rule structure, which is used in materials selection for a piston.

The selected materials are listed according to the rank, stored, and displayed in the transcript images. The most appropriate material is displayed separately as the best material and the low-level rank material is displayed according to the rank. To visualize the properties of the ranked materials, the respective rank button should be clicked. For each cycle of materials selection, the constraints and the result variables are reset to UNKNOWN. The procedure is the same for all three components. Figure 6.7 shows the result screen with selected materials for a piston.

6.6 KBS for Optimum Selection of Materials: Case Study 2

In this study, we have facilitated the calculation for optimizing material by manipulating multi-dimensional array function. MATLAB supports arrays with more than two dimensions. Multi-dimensional arrays can be numeric, character, cell, or structure arrays.

Multidimensional arrays can be used to represent multivariate data. Thus, it makes the calculation become faster, accurate, and more reliable. The user interface allows the user to input the main parameters of the problem under consideration. It also provides recommendations and explanations of how such recommendations were reached.

For selection of the optimum material for a drinking container, we defines 'function msd' that accepts inputs 'a' (weighing factor value) and 'b' (scaled property value) and returns outputs 'optimum_material_PI' to give maximum performance index which indicates the optimum material to be selected as shown in Fig. 6.8. The scale property value for the different candidate materials such as steel, aluminium, glass, polyethylene (PE) and polyethanol (PET) are calculated using standard formula as mentioned in Sect. 5.5.2. The properties of all the above materials can be seen in tutorial problem 5.14 (in Chap. 5).

The syntax 'for' is used to execute one or more MATLAB statements in a loop. Loop counter variable 'n' is initialized to value initial at the start of the first pass through the loop, and automatically increment by 1 each time through the loop. The program makes repeated passes through statements until it has incremented to

If
(Piston.ConstraintDensity > 2)
(Piston.ConstraintDensity < 2.68) And
(Piston:ConstraintStrength > 100)
(Piston:ConstraintStrength < 371) And
(Piston:ConstraintYoungModulus > 77900)
(Piston:ConstraintYoungModulus < 300000) And
(Piston:ConstraintFToughness > 3.28) And
(Piston.ConstraintHardness > 120).

Then
SetValue (CMCr.Name, 'Nicalon SiC fiber reinforced SiC matrix composites');
SetValue (CMCr.Matdensity, 2.1);
SetValue (CMCr.Matstrength, 350);
SetValue (CMCr.Matfstrength, 700).
SetValue (CMCr.Matftoughness, 8);
SetValue (CMCr.Matymodulus, 100000);
SetValue (CMCr.Cresistance, High),
SetValue (CMCr.Wresistance, High);
SetValue (CMCr.Manufac, ' Hot Pressing ');
SetValue (CMCg1:Name1,"Si3N4 reinforced with SiC continuous fiber composite");
SetValue (CMCg2.Name2,'Silicon carbide reinforced with carbon fiber composite').
SetValue (CMCg3.Name3, "Silicon oxicarbide reinforced with 3M-Brand Nextel 312
            fabric fiber composite');
SetValue (CMCg4:Name4, ' Not Available");
SetValue (CMCg1:Density1, 2.36).
SetValue (CMCg2:Density2, 2.5).
SetValue (CMCg3:Density3, 2.12);
SetValue (CMCg4.Density4, "Not Available").
SetValue (CMCg1:Strength1, 268);
SetValue (CMCg2:Strength2, 200);
SetValue (CMCg3:Strength3, 100);
SetValue (CMCg4:Strength4, 'Not Available');
SetValue (CMCg1:Fstrength1, 682);
SetValue (CMCg2:Fstrength2, 300);
SetValue (CMCg3:Fstrength3, 210);
SetValue (CMCg4:Fstrength4, "Not Available');
SetValue (CMCg3:Ftoughness3, 5);
SetValue (CMCg4.Ftoughness4, 'Not Available').
SetValue (CMCg1:Ymodulus1, 200000);
SetValue (CMCg2 Ymodulus2, 230000).
SetValue (CMCg3.Ymodulus3, 100000),
SetValue (CMCg4.Ymodulus4, "Not Available').
SetValue (CMCg1:Cresistance1, High);
SetValue (CMCg2:Cresistance2, High);
SetValue (CMCg3:Cresistance3, High);
SetValue (CMCg4.Cresistance4, "Not Available');
SetValue (CMCg1:Wresistance1, High).
SetValue (CMCg2:Wresistance2, High);
SetValue (CMCg3:Wresistance3, High);
SetValue (CMCg4:Wresistance4, "Not Available")};

Fig. 6.6 Rule structure in materials selection for a piston (Sapuan et al. 2002) (reprinted with permission of Elsevier Science)

the end value. Hence, provide the solution to find optimum performance index value and selected the best candidate material which has been shown in Fig. 6.9.

According to the optimum performance index value resulted from the KBS in MATLAB programm, aluminum has established the validity to be the best material compared to others for a drinking container. To conclude, the optimum solution for the drinking container is aluminum material.

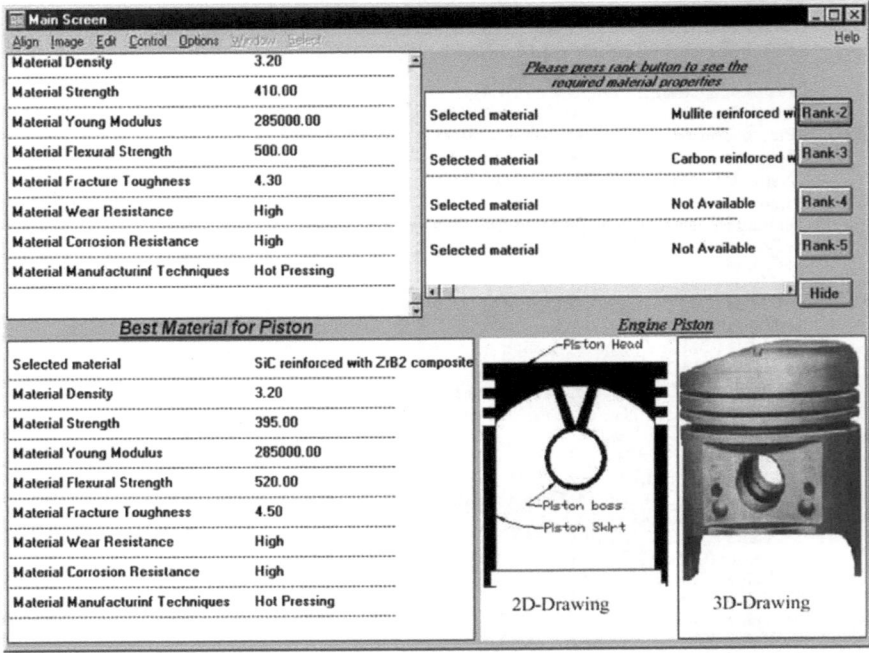

Fig. 6.7 Result screen with selected materials for a piston (Sapuan et al. 2002) (reprinted with permission of Elsevier Science)

6.7 KBS for Optimum Selection of Materials: Case Study 3

6.7.1 Analytic Hierarchy Process

Analytic hierarchy process (AHP) is an important tool used in decision making and research. AHP is a widely exploited decision making methods in cases when the decision such as the selection of alternatives and their priority is based on several criteria and sub-criteria. AHP is useful when the decision making process is complex. When the decision making process involves a number of multiple criteria where ranking is carried out according to multiple value choices, AHP splits the overall problem to solve it into as many evaluations of lesser importance as possible, while keeping at the same time their parts in the global decision. AHP method can be implemented in software tool such as Expert Choice, for individual and group decision making.

The main steps in implementing AHP are:

- Define the problem,
- Develop a hierarchical framework,

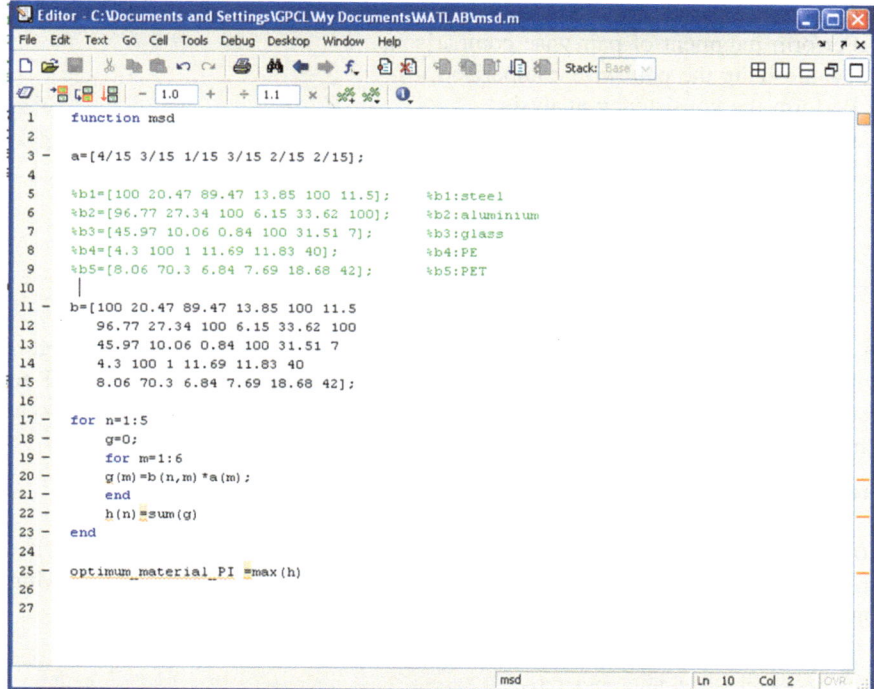

Fig. 6.8 Input variables and algorithm in expert network toolbox

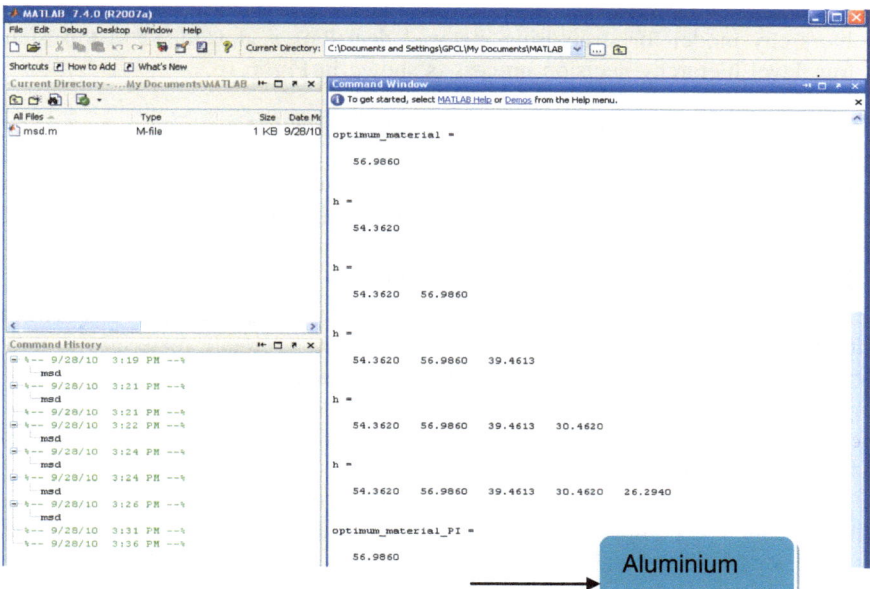

Fig. 6.9 Output results on the performance index from the system

- Construct a pair wise comparison matrix,
- Perform judgment of pair wise comparison,
- Synthesizing the pair wise comparison,
- Perform the consistency analysis,
- Develop overall priority ranking and
- Select of the best decision.

6.7.2 Materials Selection for Automotive Dashboard

In order to determine the most suitable material, AHP steps have been followed by utilizing Expert Choice 11 software. The following are the steps of using AHP for material selection of automotive dashboard (Sapuan et al. 2011) utilizing Expert Choice 11 software:

Step 1: *Define the problem*

To determine the most suitable material for automotive dashboard panel, it is important to define the problems to be solved. In materials selection the followings materials selection drivers are initially defined:

1. Density $<1,180$ kg/m^3
2. Young's modulus >2.3 GPa
3. Tensile strength >25 MPa

Step 2: *Develop a hierarchy model for material selection*

A hierarchy model for structuring material decisions is developed in this step. A four level hierarchy decision process is shown in Fig. 6.10.

Step 3: *Perform judgment of pair wise comparison matrix*

Expert Choice 11 software helps decision makers to construct pair wise comparison judgment matrix after the hierarchy model has been constructed. Pair wise comparison is done by comparing the relative importance of two selected items by using pair wise graphical comparisons provided by Expert Choice 11 software.

First of all, judgment begins with pair wise comparisons of the main criteria with respect to the overall goal of selecting the most suitable material for automotive dashboard panel. Figure 6.11 shows the judgment made on the relative importance between mechanical properties and physical properties with respect to the overall goal. The assigned value is 1.0 as shown in the Fig. 6.11. This indicates that mechanical properties are equally important with physical properties.

After the pair wise comparison at level 2 has completed, the judgment proceeds with the pair wise comparison of the sub-criteria with respect to the main criteria. Similar steps repeated on level 3.

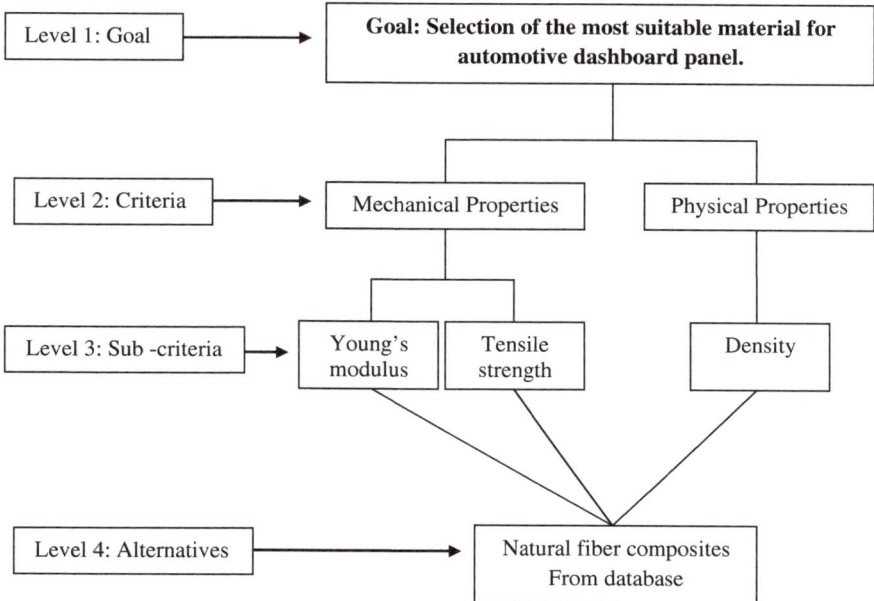

Fig. 6.10 The hierarchy model represents the criteria and sub-criteria affecting the selection of the most suitable material for automotive dashboard panel (Sapuan et al. 2011)

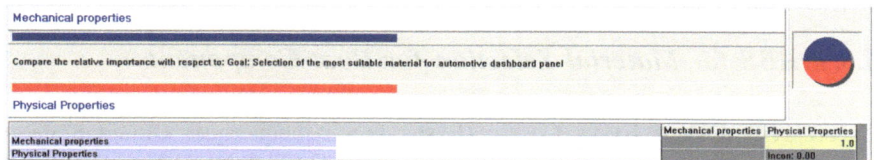

Fig. 6.11 Pair wise comparisons of the main criteria with respect to the goal (Sapuan et al. 2011)

Step 4: *Synthesizing and consistency of the pair wise comparison*

The priority vectors and the consistency ratio have to be analyzed after performing judgments on pair wise comparison. Mechanical properties and physical properties contribute the equal priority vector. As the consistency ratio (CR = 0.00) is less than 0.1, the judgments are acceptable. If consistency ratio more than 0.1, the judgments matrix are inconsistent and the judgments should be reviewed and improved in order to obtain a consistent matrix.

Step 5: *Form all pair wise comparison and priority ranking*

Step 3 and step 4 are repeated for all levels in the hierarchy model. The judgements for all levels in the hierarchy model are acceptable due to the consistency

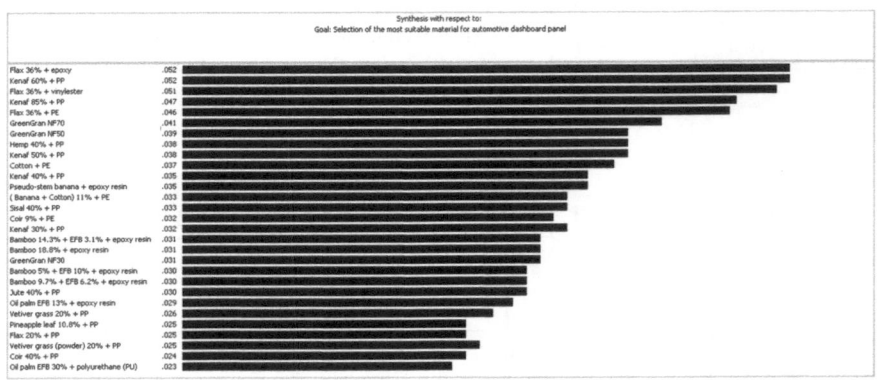

Fig. 6.12 Ranking of the materials (Sapuan et al. 2011)

ratio of less than 0.1. Finally the ranking of the most suitable material for the project purpose are shown in Fig. 6.12. The results showed that flax 36 % + epoxy and kenaf 60 % + polypropylene (PP) are the most suitable materials for automotive dashboard.

6.8 KBS for Optimum Selection of Materials: Case Study 4

6.8.1 KBS for Material Selection for Boat Components

Expert system shell Exsys Corvid (Exsys Inc, Albuquerque, United State of America) was used along with CAD software (CATIA V5R12) and material database in materials selection process. Boat components studied are rib, stanchion and beam but book only discussed the work on rib. Exsys Corvid software has database system called MetaBlock. In MetaBlock system material properties such as tensile strength, tensile modulus, flexural strength, impact strength, density, and water absorption and component price are stored.

Rule-based system
The following rules are used in material selection.
If
{attributes} < [minimum attribute specification]
Then
Score = − [attribute confident]
If
{attributes}>= [minimum attribute specification] & {attribute} <=maximum attribute specification]

Then

$$Score = \frac{[\max attribute\ spec] - \{attribute\}}{[\max attribute\ spec] - [\max attribute\ spec]} \times [attribute\ confident] \quad (6.1)$$

Then

$$Score = \frac{\{attribute\} - [\min attribute\ spec]}{[\max attribute\ spec] - [\min attribute\ spec]} \times [attribute\ confident] \quad (6.2)$$

If *{attributes} > [maximum attribute specification]*
Then
Score = [attribute confident]

Then the scoring system for each attribute is represented by score in the logic rules. There are two types of equation used to determine the score of the attributes in the Logic Block shown in Eqs. 6.1 and 6.2. Equation 6.1 is used to determine maximum value of material properties for tensile strength, tensile modulus, flexural strength, and impact strength. The closer the attribute value to the maximum attribute component design requirement, the higher the score is given to the material. Equation 6.2 is used to determine minimum value of material properties for water absorption, density, and component price. These attributes need the lowest number to gain the highest score. The closer the value to the minimum attributes value of component design requirements, the higher the score is given to the material.

A complete component design requirements for rib (tensile strength, tensile modulus, flexural strength, impact strength, density, water absorption, and component price) are set as shown in the Fig. 6.13 where the ranges of all the material properties are set depending on the specifications of the components (Fig. 6.14).

The following rules are used for the selection of materials for rib:

If

(the tensile strength of this material is >80&<160)	*and*
(the tensile modulus of this material is >1&<10)	*and*
(the flexural modulus of this material is >5&<20)	*and*
(the impact strength of this material is >100&<200)	*and*
(the density of this material is >1000&<1500)	*and*
(the water absorption of this material is >0.01&<0.1)	*and*
(the component price of this material is >500&<1000)	

Then

(This material is the most suitable for rib)

If the minimum requirement for tensile strength is set to be 80 MPa and the maximum is 160 MPa, then materials with the tensile strength lower than 80 MPa

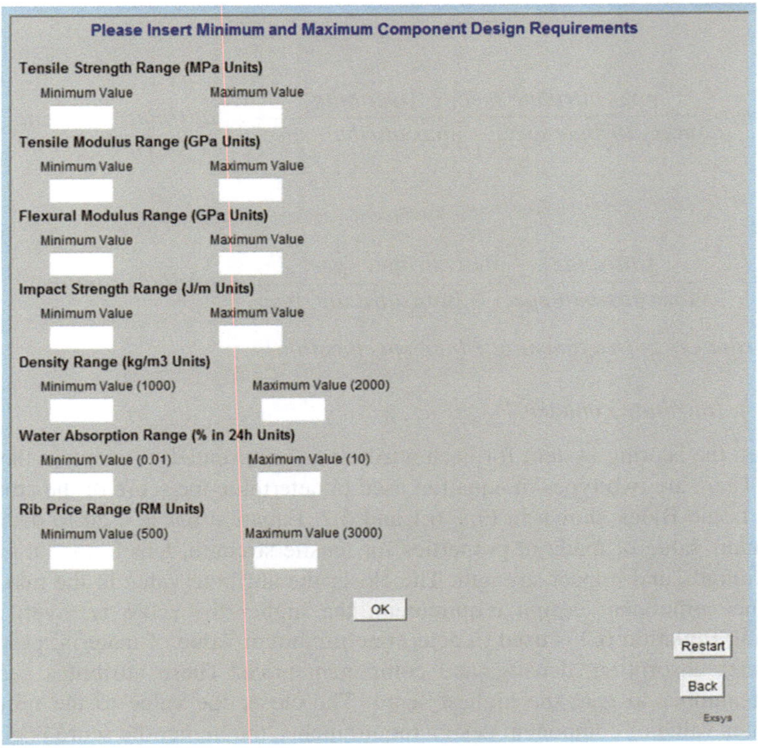

Fig. 6.13 Parameters set for rib (Fairuz et al. 2012)

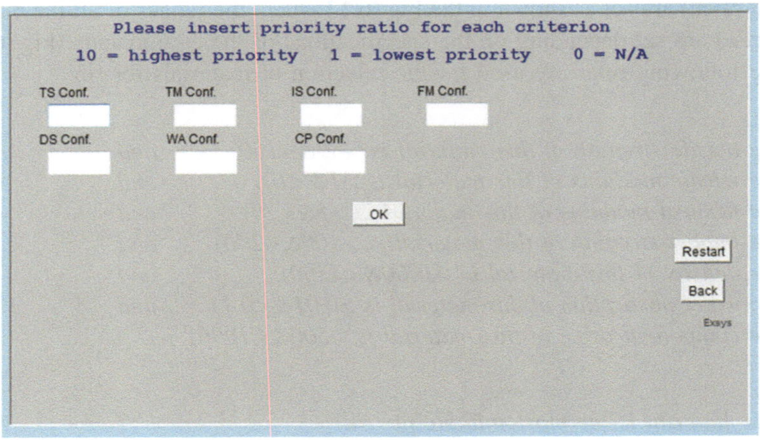

Fig. 6.14 User confident level (Fairuz et al. 2012)

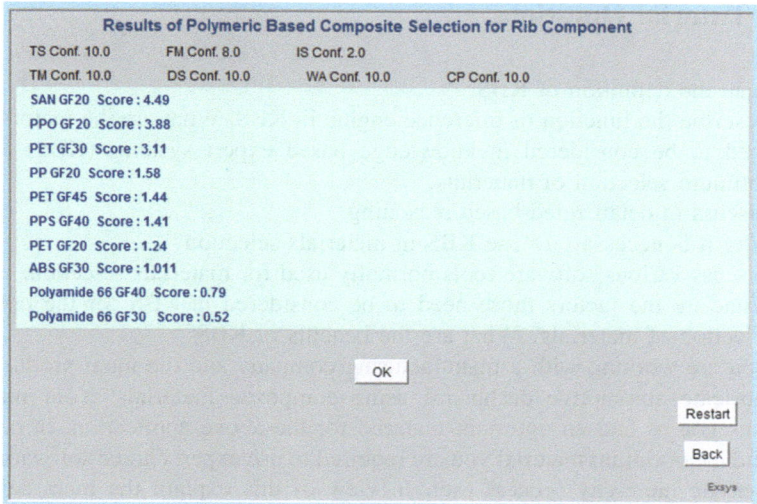

Fig. 6.15 Material selection results for rib (Fairuz et al. 2012)

will be excluded (negative value). Materials with the tensile strength higher than 160 MPa will be assigned the maximum value of 1. If a material has the tensile strength within the range, then the score of the material is assigned according to equation 1. Then the materials that satisfy the requirement or constraint can be ranked according to score from 0 to 1. Similarly, it is applicable for other requirements. On the other hand, for density, material price and water absorption, the maximum score of 1 is given to the properties that have the values lower than the minimum requirements. The selection results for rib after considering all requirements are shown in Fig. 6.15. The results show that glass fibre (20 %) reinforced styrene acrylonitrile composite is the best material for rib with the highest score of 4.49.

6.9 Summary

In this chapter KBS for materials selection is presented. Components of KBS such as rule-based reasoning, data base and inference engine are discussed. Architecture and elements in KBS are also described. Four case studies of the use of KBS for selection of materials are presented using various software packages and expert system shells such as Kappa PC, Matlab, Expert Choice and Exsys Corvid for different engineering applications.

6.10 Tutorial Questions

6.1 State the definition of KBS.

6.2 Describe the function of inference engine in KBS. What are the factors those need to be considered in knowledge based expert systems (KBS) for the optimum selection of materials.

6.3 Discuss in detail ruled-based reasoning.

6.4 Why it is necessary to use KBS in materials selection?

6.5 Discuss various software tools normally used for materials selection.

6.6 What are the factors those need to be considered in KBS for the optimum selection of materials? What are the benefits of KBS?

6.7 You are working with a manufacturing company and the main product is to fabricate automotive dashboard using composite materials. Your manager asks you to find an optimum material for the above application. In order to find the optimum material you are required to use expert choice software using analytic hierarchy process tool. In view of this explain the main select of analytic hierarchy process.

References

Durkin J (1996) Expert systems: a view of the field. IEEE Expert 11:56–63

Fairuz AM, Sapuan SM, Zainudin ES (2012) A prototype expert system for material selection of polymeric-based composites for small fishing boat components. J Food Agric Environ, accepted for publication

Gonzalez AJ, Dankel DD (1993) The engineering of knowledge-based systems theory and practice. Prentice Hall, Englewood Cliffs

Krishna MRG, Rangajanardhaa G, Hanumantha RD, Sreenivasa RD (2009) Development of hybrid model and optimization of surface roughness in electric discharge machining using artificial neural networks and genetic algorithm. J Mater Process Technol 209:1512–1520

Sapuan SM, Jacob MSD, Mustapha F, Ismail N (2002) A prototype knowledge based system of material selection for ceramic matrix composites of automotive engine components. Mater Des 23:701–708

Sapuan SM, Kho JY, Zainudin ES, Leman Z, Hambali A, Ali BAA (2011) Materials selection for natural fiber reinforced polymer composites using analytical hierarchy process. Indian J Eng Mater Sci 18:255–267

The Authors

Md. Abdul Maleque is an Assoc. Professor of Manufacturing and Materials Engineering at International Islamic University of Malaysia. He is also the Coordinator of Advanced Materials and Surface Engineering Research Unit (AMSERU). Dr Maleque is the Executive committee and life member of Malaysian Tribology Society (MYTRIBOS). He is also the member of several professional bodies and organizations. His recent work deals with materials for automotive and energy, advanced composite materials and materials selection for engineering applications. Dr Maleque served on the international advisory committee of ICAMME conference, AMPT, technical committee chairman of ICOMAST and RTC. He is on the Editorial Board of Industrial Lubrication and Tribology. He has published more than 240 technical papers in different journals and conference proceedings. He has edited 7 academic books in engineering. He was the recipient of Vice Chancellor award, UM; Dean's award, UM; IIUM Quality Research award, and IRIIE awards.

Mohd Sapuan Salit is currently a professor of composite materials in Department of Mechanical and Manufacturing Engineering, Universiti Putra Malaysia (UPM). He is also the head of Composite Technology Research Program at UPM. Professor Mohd Sapuan is the Vice President and Honorary Member of Asian Polymer Association. He is also a fellow of Plastics and Rubber Institute, Malaysia (PRIM), Institute of Materials Malaysia (IMM), and Malaysian Scientific Association (FMSA). He has successfully supervised 33 PhD and 45 MS students. To date he has authored or co-authored more than 400 journal papers and 400 seminar and conference papers. He has authored 10 books and edited 3 books in engineering. Professor Mohd Sapuan was the recipient of ISESCO Science Award, Rotary Research Award, PRIM Fellowship Award, Forest Research Institute, Malaysia Publication Award, Vice Chancellor Fellowship Prize, UPM, Alumni Award, University of Newcastle, Australia, Khwarizmi International Award and Excellence Research Award, UPM.

M. A. Maleque and M. S. Salit, *Materials Selection and Design*,
SpringerBriefs in Materials, DOI: 10.1007/978-981-4560-38-2,
© The Author(s) 2013

Index

A
Alloy steel, 23, 92, 95
Aluminium alloy, 23, 92
Analytic hierarchy process, 108

B
Bearing, 64, 67, 88, 89
Bicycle, 4, 94
Bike, 3, 4
Brake rotor, 15, 54, 82, 93
Brittle fracture, 20, 36

C
Carbon fibre, 92, 93
Carbon fibre reinforced composite, 92, 93
Carbon steel, 80, 92
Case study, 6, 9, 13, 30, 50, 65, 80, 83, 88, 103, 106, 108, 112
Cast iron, 83–86, 96
Ceramic matrix composite, 103–105
Ceramics, 1, 20, 52, 70, 74–76, 92
Composite materials, 1, 83
Corrosion, 18, 29, 53, 60, 72, 83
Cost analysis, 8, 13
Cost per unit property, 76, 78
Creep, 18, 19, 26–28, 78

D
Derating factor, 49, 50, 61
Design against fatigue, 61, 63
Design code, 48

Design consideration, 6, 58
Design phases, 3, 41
Digital logic method, 81, 87
Ductile fracture, 19

E
Elastic modulus , 75, 78
Endurance limit , 63, 64, 66, 67
Endurance ratio , 64, 66

F
Factor of safety, 49, 78, 89
Failure analysis, 29, 30, 87
Fatigue, 18, 20, 25, 58, 59, 63, 84, 89
Figure of merit, 80, 82–84
Fatigue strength, 25, 78, 89
Fracture toughness, 22, 23, 25, 92, 106

G
Glass fibres mat, 92
Graphite fibre reinforced composite, 94, 95

H
Hardenability
Hardness, 33, 36
Hierarchy model, 110, 111

I
Interface engine, 102

M. A. Maleque and M. S. Salit, *Materials Selection and Design*,
SpringerBriefs in Materials, DOI: 10.1007/978-981-4560-38-2,
© The Author(s) 2013